T0353573

A Theology of International Development

Religion and development have been intertwined since development's beginnings, yet faith-based aid and development agencies consistently fail to consider how their theology and practice intersect. This book offers a Christian theology of development, with practical solutions to bridge the gap and return to truly faith-based policies and practices.

Development aims to raise the living standard of the world's poor, mainly through small-scale projects that increase economic growth. A theology of liberation provided a critique to development practice, but a specific theology of development is still lacking, and many faith-based aid agencies have failed to adapt their practice. In applying theological thinking to development, the author argues that aid agencies need to address the entrenchment of unequal power relations, and embrace a holistic notion of development, defined by the needs of those most marginalized, instead of by a focus on economic growth. Development organisations need to consider the distinction between charity and justice, and to empower people in the global South, paying particular attention to the intersections of race, class, sexuality, religion, and the environment.

Overall this book is a powerful call to upend development practice as it currently exists and to return faith-based organizations to following Christian practices. It will be an important read for religion and development researchers, practitioners, and students.

Thia Cooper is Professor of Religion and Latin American, Latinx, and Caribbean Studies at Gustavus Adolphus College, USA.

Routledge Research in Religion and Development

Series Editors:
Matthew Clarke, Deakin University, Australia
Emma Tomalin, University of Leeds, UK
Nathan Loewen, University of Alabama, USA

Editorial board:
Carole Rakodi, University of Birmingham, UK
Gurharpal Singh, School of Oriental and African Studies,
University of London, UK
Jörg Haustein, School of Oriental and African Studies,
University of London, UK
Christopher Duncanson-Hales, Saint Paul University, Canada

The *Routledge Research in Religion and Development* series focuses on the diverse ways in which religious values, teachings and practices interact with international development.

While religious traditions and faith-based movements have long served as forces for social innovation, it has only been within the last ten years that researchers have begun to seriously explore the religious dimensions of international development. However, recognising and analysing the role of religion in the development domain is vital for a nuanced understanding of this field. This interdisciplinary series examines the intersection between these two areas, focusing on a range of contexts and religious traditions.

Secular and Religious Dynamics in Humanitarian Response
Olivia J. Wilkinson

Nigerian Pentecostalism and Development
Richard Burgess

A Theology of International Development
Thia Cooper

Muslim Women in the Economy
Development, Faith and Globalisation
Edited by Shamim Samani and Dora Marinova

A Theology of International Development

Thia Cooper

Routledge
Taylor & Francis Group

LONDON AND NEW YORK

First published 2020
by Routledge
2 Park Square, Milton Park, Abingdon, Oxon OX14 4RN

and by Routledge
52 Vanderbilt Avenue, New York, NY 10017

Routledge is an imprint of the Taylor & Francis Group, an informa business

British Library Cataloguing-in-Publication Data
A catalogue record for this book is available from the British Library

Library of Congress Cataloging-in-Publication Data
Names: Cooper, Thia, author.
Title: A theology of international development / Thia Cooper.
Description: Abingdon, Oxon; New York, NY: Routledge, 2020. |
Series: Routledge research in religion and development |
Includes bibliographical references and index. |
Identifiers: LCCN 2019054013 | ISBN 9780367332013 (hardback) |
ISBN 9780429318412 (ebook)
Subjects: LCSH: Economic development—Religious aspects—
Christianity. | Economic development—Developing countries. |
Poverty—International cooperation. | Religious institutions—
Developing countries. | Christianity.
Classification: LCC BR115.U6 C66 2020 | DDC 261.8/5—dc23
LC record available at https://lccn.loc.gov/2019054013

ISBN: 978-0-367-33201-3 (hbk)
ISBN: 978-0-429-31841-2 (ebk)

Typeset in Times New Roman
by codeMantra

This book is dedicated to my mother, who gave all that she had and often went without, so that others could live abundant life

Contents

Acknowledgments

First, I want to thank Helena Hurd at Routledge for suggesting I consider writing a book in this area. Second, I want to thank the reviewers for their insightful comments. Third, I want to thank my fellow students in the MTh in Theology, Culture, and Development at the University of Edinburgh (1999), way too many years ago now. Without each of you and my fellow PhD colleagues, I would not be where I am in my thought process today. Fourth, I want to acknowledge the huge influence Professor Marcella Althaus-Reid had in my life. I still miss her every day. Fifth, I want to thank my online writing group at Academic Ladder for their constant support and Anna LoFaro for all her help this summer. And finally, I want to thank my mother Cynthia, for moving to be near me in her final months of life, Raoul, and my friends for putting up with me during this less than restful sabbatical year.

Introduction

"I work with real believers," said the director of the Jordan Health Aid Society (JHAS) in Amman. In 2018, I and several colleagues traveled to Jordan meeting with politicians, nongovernmental organization (NGO) directors, health care workers, tribal leaders, and many other people to learn more about how Jordan navigates regional politics and in particular Jordanian work with refugees, most recently Syrian but also Iraqi and Palestinian refugees.[1] JHAS provides health care to refugees and poor Jordanians, trying to bridge the gap between NGOs and government provision of health care. A colleague asked the director how the diverse religions of the people involved affected his work. "A real believer is not someone who wants to further their own career or to make money. A real believer works to end poverty,[2] with those most at risk, often risking their own lives." He stated that a real believer can come from any religious tradition. Real believers are committed to ending suffering. They focus on enabling others to take over the work they begin. They want to empower people to sustain themselves. His description mirrors what I think of as a theology[3] of development, as I will elaborate in this book.

"Development"[4] practice attempts to improve the lives of the poor. The question is: does it? This book argues that while some shifts within development work do prioritize the poor, current development practices still work against the Christian notion of "life abundant"[5] and Christian theology has failed to articulate a theology of development that leads to coherent practice.

The extreme poverty in our world today is a scandal, yet most of us fail to recognize it as such. Millions of people die prematurely each year, many due to the effects of poverty. Harsh Mander, an Indian scholar and activist, in his excellent book *Looking Away* states that more than 300,000 people die in India from avoidable causes each year.

> There is little substantive difference between genocide and simply allowing poor people to die. ... The reason why the preventable deaths of these many millions year after year is not "considered exceptional, a tragedy and a disgrace," according to [Akhil] Gupta, is the *normalization* of poverty
>
> (Mander 2015, p. 3)

While we are horrified at genocide, we accept deaths from poverty as a normal part of our world.

While official policies of development[6] began post-World War II with the attempt to bring to other countries the restoration that worked in Europe, for many parts of the world, particularly Asian and African countries, development was a continuation of colonialism and missionary work in another guise. The history of the Department for International Development (DfID) in the UK shows this link between colonialism and development. DfID began in 1929 under the UK Colonial Development Act, later becoming the Ministry for Overseas Development, and finally DfID in 1997. Under colonialism, the colonists and the home countries saw the colonies as a source of raw materials, including free labor. Policies of colonialism usually emphasized the colonizing country's progress and did not promote progress in the colonies, except for the colonists. To what extent would development enable countries to engage in progress or prevent people from making their own decisions?

As U.S. President Truman initially articulated in 1949, development would bring "underdeveloped areas" up to the level of the USA through capitalism[7] and democracy with a sharing of "technical knowledge." Within a decade, partly due to the Cold War, this focus had narrowed to capitalism. The USA and Western Europe supported dictatorships in many countries, fearing a move to communism if left to democratic governments. Capitalism was assumed to be the mechanism best suited for economic growth, benefitting all countries. Poverty was an economic problem, requiring an economic solution. This poverty was a lack of money and that lack was consistently relative as growth was key to the system. In effect, a country never reaches full development as individuals and nations have unlimited desires, according to capitalism. Development then, as a subset of capitalist economics, aimed to raise the living standard of the world's poor through economic growth. This understanding narrowed the original concept of development as change and progress including all realms of society. It did, however, keep the assumption that the North[8] was a model for the South to follow. Development today tends to retain this assumption as well as the focus on economics.

Since the notion of development emerged, many Christians[9] have supported development.[10] In many cases, it continued under the label of mission work, particularly in education and health care. Christian faith-based aid and development organizations (FBDOs)[11] assume Christianity supports development practice and these organizations rely on the support of Christian churches and Christians in the global North. However, the history of development and liberation show that while there is a theology of liberation, there is not a coherent theology of development. Those hoping for a theology of development started with questions and stopped before offering a theology because the questions led to a critique of development as practiced. Many FBDOs co-opted the language of liberation without changing practice.

Why do we need a theology of development?

Romy Tiongco, a former Christian Aid staff member and Filipino Roman Catholic theologian, asked this question at the beginning of his book: *Doing Theology and Development: Meeting the Challenge of Poverty.* "Development is concerned with the immediate, practical challenge of tackling world poverty. ... Theology seems too abstract, too removed, more concerned with the hereafter than the here and now" (White and Tiongco 1997, p. 1). In fact, this charge has been leveled at traditional theology by many theologians, including myself. This book aims to link Christian theology with development practice to enable FBDOs to put faith into practice in a way that helps rather than harms the most marginalized. Theology cannot remain abstract and development cannot ignore ethical issues.

First, many Christian FBDOs are involved in development practices around the world. Second, it is not always clear how the Christian aspect of these agencies differentiates their practice from secular agencies. Third, people involved in development work in the North often do not see the point of integrating theology and development. As Christianity became less prominent in the global North during the twentieth century, it was also less important to those "doing" development. However, many people experiencing development still viewed Christianity (and religion in general) as important and so I analyze it from this direction. Religion is surging in the global South and is resurging in some parts of the North. Fourth, people living in many Southern countries incorporate their faith with other aspects of their lives. Their development practice includes spiritual elements, and their theology includes economic, political, and other elements. Finally, the majority of Christians in the world live in areas considered to need development: Latin America, Africa, and Asia, not in the "West." A 2011 Pew Research Center Report found that 61% of the world's Christians lived in the global South.[12] (p. 13) Hence, how Christianity can support "good change" is critical for the Christian faith. Could Christian FBDOs be prophetic?

While Christian FBDOs may be trusted by Christians in developing countries, more so than secular organizations, people in other religious traditions also tend to trust faith-based organizations more than secular ones. Bob Mitchell, CEO of Anglican Overseas Aid in Australia, noted: "Women from one program stated that in doing development it is 'important to have faith. Christian or Muslim doesn't matter'" (Mitchell 2017, p. 47). People tend to assume that faith-based organizations have a coherent ethical rationale for their practices. Unfortunately, this rationale often does not exist or is incoherent.

I write this book, first, because there is no book articulating a Christian theology of development and practice.[13] While several books attend to aspects of theology, none fully develops a theology and practice. Second, FBDOs have failed to coherently integrate their theology and practice. As noted in my previous book, *Controversies in Political Theology* (Cooper 2007), there is a gap between the stated theology and the effects of practices. While the previous book traced this gap through the history of development, contrasting it with the theology and practice of liberation, which states that Christianity should free people rather than oppress them, this current book offers a theology of development, along with ways to implement ethical policies and practices. Third, development studies literature has tended to treat development as a secular concept. However, Christianity and development have been intertwined since development's beginnings. It is important to understand how the Christian tradition supports and critiques development and how those experience development critique Christian tradition and practices.[14] Many of the students I encounter in the global North want to do good in the world but rarely interrogate how their ethics and practices intersect; this book aims to respond to this lack. The book articulates why strands of the Christian tradition focus on the poorest and how, according to that tradition, Christians can help rather than harm others, through an understanding of power, justice, and the marginalized.

Christianity does not stand outside of history; it is part of history. It is in the reality of history that people relate to God and each other. At the same time, "God's kingdom,"[15] which I will refer to as the new heaven and new earth, is not a solely spiritual concept; it is a real-world concept too, which requires working in the here and now, as I will explore. Christians are working either toward better relations with others and God or away from them. Let's start with a basic concept which I will develop further. The Gospels tell us Jesus when speaking of the judgment of the nations said: "Truly, I tell you, just as you did it to one of the least of these who are members of my family, you did it to me" (Matthew 25:50). Jesus is the other person with whom we are in relation. Are we treating Jesus justly or unjustly? Building good relationships with others is important here. If we build just relationships with others, we are building a just relationship with God. If we build unjust relationships with others, we are building an unjust relationship with God, separating ourselves from God.

A theology of development will articulate how the Christian tradition supports empowerment, justice, and walking with the marginalized. Theology goes beyond what is happening to argue for what should happen. In particular, aid agencies could address the entrenchment of unequal power relations, the distinction between charity and justice, and what it means to prioritize those most marginalized. The first section of this book addresses those three issues from the perspectives of Christian theology and ethics. Theologically, I develop the notion of power as empowerment rather than "power over." I unpack the false theological distinction made today between charity and justice, which has led Christians to work toward charity rather than justice. Finally, I articulate the concept of walking with the marginalized, again based on biblical scholarship and tradition.

The second section of this book articulates the three shifts in practice that could result from thinking theologically about development. First, the organizations could aim to reduce the power of the wealthy in the global North, while empowering people in the global South. Aid agencies need to address the entrenchment of unequal power relations by shifting their modes of practice. Second, the organizations could end the focus on economic growth within the capitalist system as justification for policies and return to a notion of economic development defined by those most marginalized. This includes ending both poverty and accumulation of wealth. Third, the organizations could attend to the intersections of race, class, sexuality, religion, the environment, etc., which intersect to marginalize people within and between communities. Making these changes would enable the poorest to make decisions for themselves, upending development as "normally" practiced.

This book is important for those working in the development field and those who want to alleviate poverty for several reasons, not least of which is articulating a practical theology of development. This book also attends to how theology and practice intersect. Further, the concepts of empowerment and intersectionality are not yet main-streamed within development studies.[16] I will introduce a theology of development from the perspective of the majority of Christians: those Christians live outside the USA and Europe and live with the effects of development. Most attempts at a theology of development started with questions from the top-down and stagnated, as we will see. They ask: how can "we" help to improve the lives of others? Instead, the question should begin with the marginalized themselves, who form the majority of the world's Christians. What do the marginalized want in order to be able to live life abundant?

Alone, the changes to the work of Christian development practitioners and the organizations themselves will only be a small piece of the field of development. Governments, international organizations such as the World Bank, the World Trade Organization (WTO) and the International Monetary Fund (IMF), alongside many secular aid and development organizations all play a role. What this book suggests is that by overturning the practice in this small sector, there can be larger scale results, particularly in the realm of advocacy. FBDOs could and should advocate in these broader organizations and encourage their supporters and the churches with which they are connected to do the same. FBDOs should reflect the demands of the most marginalized to the wider world, amplifying their voices.

How will I "do" theology?

The approach to theology matters; no theology is neutral; our experiences shape our beliefs and practices. Further, by avoiding a decision, I am already supporting the status quo. My mom raised me on her own in a poor U.S. neighborhood. We attended a Christian Pentecostal church, where I was heartened by the help the church gave individuals in need but disheartened by the focus on getting to the afterlife (through salvation) where suffering would finally be relieved. What about working to end suffering now? At university, I focused on international relations and development and found in this field, there was a lack of attention to the role that religion plays in influencing the daily lives of individuals and structures. In graduate school, I explored deeply the intersection of faith and life, becoming a liberation theologian. Though I completed an MSc in Development Studies, I rejected development as a harmful top-down approach and since my PhD in theology, I have written from the perspective of liberation theologies on topics from sex, to economic poverty, to reproductive justice. The field of development, on the other hand, persists and many FBDOs continue to persist within it. Hence, the theology in this book will come from the perspective of Christians experiencing development to find a middle ground whereby Christian FBDOs can practice a form of development consistent with the Christian tradition.

My method comes from liberation theology and is called the hermeneutical circle, a circle of action and reflection (Figure I.1). This theology emerges in community. I do not develop this theology alone. As Kwame Bediako, a Ghanaian theologian, noted, in the global South, though individual theologians still publish "the theological activity itself is seen as a shared activity emerging out of a shared experience. ... The explicit claim is that this is the 'theology of the people.'" (Bediako 2004, pp. 112–3) It is theology that emerges from community and with which I agree.[17]

The first step in the circle is reading reality from the community's experience. Paulo Freire calls this conscientization.[18] One's situation has to be examined critically: in this case, the experience of development. We need to become deeply aware of and analyze how communities experience development.

The next step is the dialectic of scripture and reality, which I extend to scripture, tradition and reality. (Althaus-Reid 2000, p. 389) With scripture and tradition, I begin with Jesus and the earliest church. In doing theology, "a theology that is not nourished by walking Jesus' own path loses its bearings" (Gutiérrez 2013, p. 151). The reality

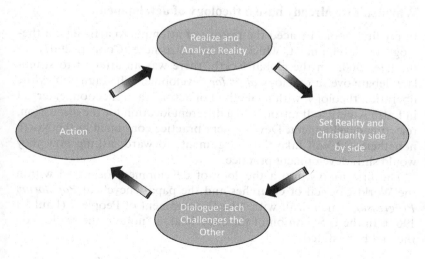

Figure I.1 The Hermeneutical Circle.

discerned is the starting point for engaging with Christianity. The realities of development and theology are set alongside each other in order to dialogue.

The third step is this dialogue, where the experience of development and Christianity talk together and challenge each other. What does Christianity have to say about our experience of development? What does our experience of development have to say about Christianity? There is continual intertwining of the reading of reality and Christianity, which leads to the final step.

The last step is action. Reading reality and Christianity together should lead to a change in action. This is praxis, action which has been reflected upon. There will not be a systematization of the results, as is often assumed with dogma (right rules), because the process continues. Praxis continues to progress as reality changes. I would hope that a few years from now, there would be another book challenging the themes I have laid out here, reflecting further on the themes from experience.

Finally, crucial to the hermeneutical circle is "suspicion." One does not examine theology to try to match it to an already defined reality. Further, one does not examine reality to try to match it to a previously defined theology. Instead, the two are integrated, speak with each other, and teach each other. Those participating in the hermeneutical circle need to be open to having perspectives shifted. As such, this book aims to offer starting points to further theological praxis.

Why don't we already have a theology of development?

In my first book, I traced the history of attempts to articulate a theology of development, which I summarize here (Cooper, 2007). In the late 1960s, in the global North, there was an attempt to stimulate debate over a theology *of* or *for* development. It stagnated, while liberation theology, with its method of action and reflection, emerged in Latin America. It introduced a different location for the discussion, rejecting development. Development practice continued in the North nonetheless. I will take these arguments forward, asking how they would shift development practice.

The first forays into a theology of development occurred within the World Council of Churches and the papal encyclical *Populorum Progressio,* which dealt with "the Development of Peoples" (Paul VI 1967). In the U.S. edition of the encyclical, a "note to the reader" on the first page stated:

this individualistic system of capitalism so strongly condemned by Popes Leo XII, Pius XI, Pius XII, John XXIII, and Paul VI involves a type of calloused exploitation which certainly is not descriptive of the prevailing business practices of the United States in 1967. ... Such warnings are indeed timely, since exploitive capitalism does characterize the economic systems of many of the poorer nations of the world.

Before reading the text, the reader was informed that any critique was not meant for the USA, only poorer countries. Thus, where the Pope was critical of capitalism, it was applicable to underdeveloped countries. One should not mistakenly think that the global North might need to change too. Even when a theology started to be articulated, pieces of it were ignored.

The World Council of Churches also approached a theology of development, first by holding a World Conference on Church and Society in 1967, called "Christians in the Technical and Social Revolutions of our Time." A theology of development did not emerge from this conference but a theology of liberation emerged soon after. Together, Catholics and Protestants created the Joint Committee on Society, Development and Peace (SODEPAX) in 1968 to continue work on the themes.[19] In 1974, Alistair Kee edited a *Reader in Political Theology*, which contained a section of chapters on theology and development.

The first characteristic of these writings on theology and development is an acceptance of the concept of development. It is not unquestioning or uncritical; nevertheless development is accepted *apriori*. The second characteristic is the assumption that theology should reflect on the process of development. Theology could provide a motivation for development, could analyze the goals of development, and the method used in development. (Loffler 1974) However, the practice of development was not to consider or change theological concepts itself. The experience of those being developed would not challenge theology or development practice. As we know, the field of development often ignored theology, although theologians still reflected on aspects of development.

In essence, these groups supported development and then asked how theology could too. Yet, the emerging theology was either critical or suggested areas of focus not currently part of development. Thus, it faltered because its aim was to articulate how development was positive and could be supported by Christian principles. Initially, four questions emerged: (1) Is development salvation? (2) What should the North do with its riches? (3) What do the new heaven and the new earth mean for economic and political structures? (4) Should humanity be at the center of development?[20] Each of these questions led to the beginnings of a critique of development. However, the theologians were aiming to find a way to encourage Christians to become engaged in development, rather than to critique it or change it, so the conversation stopped at the questions and the theologians either continued their support for development, moved to overtly supporting capitalism and later globalization,[21] or worked on the emerging liberation theologies, rejecting development.

I want to take this theological discussion forward and suggest a theology of development, critique and all. What would development based on Christian principles look like? First, by considering development in the light of salvation, what development could be becomes clearer. By connecting salvation and development, "I mean the transfiguration and transformation, in the broadest possible meaning, of the world" (Nissiotis 1971, p. 153). Salvation is broader than the individual, in this view, it is for the world and all the people within. If we think of development in the sacred sense of salvation, what should it look like? How can each human being on the planet live life abundant in right relationship with God and others? It certainly means that development cannot be limited to the economic realm.

In response to the question of what the North should do with its riches, the theology led to the idea that just as poverty should be reduced, so too should the riches of the North. *Gaudium et Spes* (1965) argued, "If a person is in extreme necessity, he has the right to take from the riches of others what he himself needs" (para. 69). We will consider the concept of ownership in the chapter on charity and justice. Rather than follow this perspective, some theologians in the North moved to supporting capitalism and the accumulation of wealth. In this argument, the North was to use its riches wisely: consuming and stewarding, since the North leads the global economy.[22] The questions of power and inequality were set aside. The poor should be helped. However, the rich did not need to change. I'll articulate throughout this book how development cannot simply focus on improving the lives of the poor without attending to excessive wealth and power.

Further, in terms of the new heaven and new earth, initially the theology led to the suggestion that "The Christian is therefore called to speak a radical 'No' – and to act accordingly – to the structures of power which perpetuate and strengthen the status quo at the cost of justice to those who are its victims" (World Conference 1966, p. 200). Nothing should be accepted without question, including development. Again, we take this notion forward asking what development would look like if this structural critique was applied. In contrast, as theologies of capitalism emerged, the main argument became to do the best we can within this globalized capitalist system. Rather than accept the status quo, I ask what we should do, suggesting alternatives. FBDOS could support the creation of alternative institutions not focused on economic growth and power as domination, as noted in later chapters.

Finally, in response to the question of whether humanity should be at the center of development, the suggested answer seemed to be yes. "Development cannot be limited to mere economic growth. In order to be authentic, it must be complete: integral, that is, it has to promote the good of every man and of the whole man" (Paul VI 1967, pp. 13–14). When assessing development, the good of the human being should be at the center, a concept broader than economics and a challenge to capitalism. Again, as theologies supporting capitalism and globalization emerged, the focus narrowed from humanity to the individual within the market. This book returns the focus to the poorest and most marginalized and examines what it means to be fully human. What issues need to be addressed for humans to live life abundant?

At the beginnings of the conversation over theology and development, these questions led to a theology of liberation. "The term liberation avoids the pejorative connotations which burden the term development" (Gutiérrez in the Consultation on Theology and Development 1970, p. 125). Rather than delve into the emergence of the theology of liberation, I will take these questions forward and see if the term "development" and its practices can be rescued from its pejorativeness, analyzing practice in terms of power, justice, and the marginalized.

How do people criticize development?

There are numerous critiques of development by those experiencing it. Let me cite just two here to set our scene. The first emphasizes the notion of inferiority underlying much development practice; the other emphasizes the links to colonialism and neoliberalism.

Critiquing the notion of development as progress toward the level of the global North, Esther Mombo, Kenyan professor and theologian, states: "While this version of development is no longer mainstream, it continues to linger in the models of some theorists. ... Development thus became an alienating and humiliating process for the people who helplessly remain undeveloped" (Mombo 2009, p. 219). Because countries were considered to be along a spectrum from "backward" to "advanced," development became a disempowering and embarrassing experience. Paul Farmer, U.S. medical anthropologist and doctor, echoes this assessment with regard to provision of health care: "Developmentalism not only erases the historical creation of poverty but also implies that development is necessarily a linear process: progress will inevitably occur if the right steps are followed" (Farmer 2013, p. 57). In the health sector,

> one sees the impact of "developmental" thinking ... in the "health transition model." In this view, societies as they develop are making their way toward that great transition, when deaths will no longer be caused by infections such as tuberculosis but will occur much later and be caused by heat disease and cancer. But this model masks interclass differences *within* a particular country. For the poor, wherever they live, there is, often enough, no health transition.
>
> (Farmer 2013, p. 58)

Critics argue that development is not a linear progress, with some countries lagging behind. There are varying rates of "progress" within and between countries. Further, people understand progress in different ways. The notion that some countries are developed masks the fact that not everyone within those countries has the same access to life-sustaining resources.

Development is also critiqued as an updated form of colonialism, continuing the condescension of the North toward the South and the prioritization of the North over the South. May Ngo, a Cambodian anthropologist, argues,

> The IFIs [International Financial Institutions] utilize the concept of good governance as a "bridging concept" to articulate a new relationship between their controversial economic policies and human rights and development (Anghie 2007). By reformulating the dynamic of difference as one of being between "good governance" and "bad governance," the IFIs are able to "cast the restructuring of Third World political and public institutions as a necessary step toward realizing global human rights" (Mahmood 2015, p. 92).
>
> (Ngo 2018, pp. 10–11)

Northern countries and institutions moved from seeing the South as behind the curve culturally, to economically, and now to politically. In each of these realms, the North argues that the South needs to be "helped," although benefits should also accrue to the North in the form of economic growth and safety and security. Underdeveloped countries never achieve equality with developed countries. First, the goalposts move. Second, developed countries continue to accumulate. So it is impossible to "catch up."

One particular critique of NGOs is that they support, either overtly or inadvertently, global capitalism, neoliberalism, and the Northern "addiction to help." NGOs fill in the spaces the economies and governments avoid, enabling economies and governments to continue to avoid the gaps.

> Even when unintentional, actions by faith-based actors can carry a wider political force particularly within a context where faith-based organisations are often filling in a role left by the state, thereby engaging in action which may challenge conceptions of who can legitimately be a beneficiary of care and services. This is especially true of organisations working with those deemed outside of the State's [mandate].
>
> (Ngo 2018, p. 5)

NGOs, in working with the poor, often provide services that national and regional governments do not provide in the current global situation. Particularly if NGOs do not work with governments, they are supporting and enabling, rather than struggling against the dominant notion that the economy should take priority over government provision of services and creating a third space, known as "civil society,"[23] which is now expected to fill in where the economy and the government cannot.

With these critiques in mind, let us also note a positive aspect from one UK development practitioner and theorist, Robert Chambers.

> It is thrilling too because meanings of "development" continuously evolve and diversify. ... The 17 Sustainable Development Goals (SDGs) apply everywhere. All countries have signed up to them. The old dichotomies and mindsets of donor-recipient, North-South, developed-developing are superseded.
>
> (Chambers 2017, p. xi)

Is his understanding of development as having expanded and now being for everyone possible in practice and could Christian FBDOs help to achieve it?

Notes

1 Jordan is a country with about 10,500,000 people. It is 89th in the world according to GDP. Yet it has hosted about 1.5 million Syrian refugees, roughly 500,000 Iraqi refugees, and continues to host Palestinian refugees.

Palestinians number about 2,000,000 in Jordan. The CIA World Fact book lists 69.3% of the population as Jordanian, 13.3% as Syrian, 6.7% as Palestinian and 1.4% as Iraqi. See www.cia.gov/library/publications/the-world-factbook/geos/jo.html, https://www.unhcr.org/en-us/subsites/iraqcrisis/47626a232/iraqis-jordan-number-characteristics.html

2 Poverty is discussed and defined in the second chapter. Poverty is commonly defined with relation to economics, although this definition has broadened within development studies. Some literature has added the concept of well-being. One trend asks how people themselves define poverty and what it would take to lift them out of poverty. This is the trend I prioritize in this book.

3 Theology can be defined as faith seeking understanding, a definition by Anselm, in this case an understanding of poverty. Theology is faith in action. It is not simply thinking about what should be or what truth is but includes how to practice faith. Traditionally, theology was separated from ethics: theology was the thinking and ethics was the doing. However, this is a false distinction. If I do something different to my thinking, the doing shows much more about my beliefs than my thoughts. Theology and ethics are not separate.

4 "A quote from one of the interviewees referring to the moment when he was required to leave the colonial service reveals the perceived link between the work carried out in the late colonial period and the work of post-war development. 'And I thought, right, if I can no longer do this job and work out here the next best thing is to be working for the development of Kenya in the development field – after all, it is the same thing. ... But what's the point of chasing a dwindling colonial empire around – let's get back and get our teeth into something that will be important – helping Third World Countries'" (Kothari 2005, p. 56).

5 John 10:10 (NRSV) states, "I came that they may have life, and have it abundantly."

6 Development is a contested term, commonly defined as economic growth with poverty reduction. It tends to be defined according to what people want to occur rather than what is actually occurring in practice. Gilbert Rist, a professor of development, described such definitions as wish lists. Rist's definition then is: "'Development' consists of a set of practices, sometimes appearing to conflict with one another, which require- for the reproduction of society- the general transformation and destruction of the natural environment and of social relations. Its aim is to increase the production of commodities (goods and services) geared, by way of exchange, to effective demand" (1997, p. 13). The verb 'to develop' does not have this negative meaning historically. It can also mean "to cause to unfold gradually" or even "to become gradually manifest." Development then would be "the act, process or result of" this unfolding gradually.

7 Capitalism can be defined as "an economic system in which wealth is owned by private individuals or businesses and goods produced for exchange, according to the dictates of the market" (Heywood 1992, p. 311).

8 Throughout this book I will use "North" and global North to refer to wealthy countries, which are geographically clustered in the northern hemisphere, with a few exceptions like Australia. I will use "South" and global South to refer to poorer countries, which are geographically clustered in the southern hemisphere.

9 People of other traditions have also argued for and against development from within their traditions. I do not detail them in this book, as I am not an expert in these traditions. However, I will refer occasionally in footnotes to material that could be relevant for interreligious discussion on development.

10 "A research report in 1953 found that 90 per cent of the post-war [WWII] relief was provided by religious agencies (Ferris, 2005)" (Hoffstaedter And Tittensor 2013, p. 402).

11 Here, I use FBDOs to mean faith-based aid and development organizations. There are a variety of definitions of FBDOs. See Tomalin (2013, pp. 216–221) for details. FBDOs include Christian Aid, CAFOD, World Vision, Compassion International, Lutheran World Relief, and Habitat for Humanity, as well as any local organization working on community issues from a faith base. For the purposes of this book, I address any organization involved in aid and development work that purports to work from within the Christian tradition. As I write, however, I address mainly Christian organizations from the North working in or with partners from the South. Terms used also include "religious NGOs," common in the UN. (Occhipinti 2015, p. 335).

For example, Christian Aid is one FBDO. CA is a UK-based relief and development agency, which funds agencies indigenous to the overseas countries, rather than managing its own overseas projects. From the broadest perspective, it can be seen in the middle of two arms. On one side it has been set up by these UK churches. On the other side, it funds and interacts with many secular and faith-based partners in countries around the world. CA has to balance these two arms within the organization. While these two often seem to be in tension, this book offers a way for this organization and other Christian-based development organizations to hold the two together theologically.

FBDOs fall under the broader rubric of faith-based organizations (FBOs).

> Gerard Clarke and Michael Jennings…offer a broader definition of an FBO as "any organization that derives inspiration and guidance for its activities from the teachings and principles of the faith or from a particular interpretation or school of thought within the faith" (2008, 6)
> (Occhipinti 2015, p. 335)

For example, a church may engage in development work without being a formal development organization.

12 For example, in terms of the largest numbers of Christians, the top ten countries as of 2015 were (1) USA; (2) Brazil; (3) Mexico; (4) Russia; (5) Philippines; (6) Nigeria; (7) China; (8) Democratic Republic of Congo; (9) Germany; (10) Ethiopia. Together, Brazil, Mexico, the Philippines, Nigeria, China, the Democratic Republic of Congo and Ethiopia have 29% of the world's Christians.

13 There are several books that attend to the subject in part, for example: Stan Chu Ilo's *The Church and Development in Africa* (2011), Catherine Loy's *Development Beyond the Secular* (2017), Bob Mitchell's *Faith-Based Development* (2017), and Sarah White and Romy Tiongco's *Doing Theology and Development* (1997).

14 For example, an often cited study is Kurt ver Beek's examination of three development studies journals between 1982 and 1998. (ver Beek 2000)

During this time there were few articles that mentioned religion or spirituality. In contrast, however, if you examine religious studies and theology journals during this time, aspects of development studies are regularly referred to. A cursory examination of the ATLA database showed more than 150 before I paused my search. Religious studies considers development, although development studies may have tried to ignore religion.

15 Often, this concept is characterized as the "kingdom of God." However, this terminology has a host of problems, regarding the power hierarchy of a kingdom and the male-centered nature of it. See, for example, Soelle, Chapter 12. I use the terminology of a new heaven and new earth from II Peter 3:13 "But, in accordance with his promise, we wait for new heavens and a new earth, where righteousness is at home." See also Isaiah 65:17 and 66:22.

16 Intersectionality can be defined as the ways in which different parts of a person's identity intersect with each other and are either privileged by or discriminated against by society. For example, I am a white woman. I have privilege in my whiteness and experience discrimination in my femaleness. If I was a poor lesbian woman, I would have more intersections of oppression.

17 As part of this articulation, I have tried to ensure the majority of my sources are women and people of color, mainly from the global South.

18 Conscientization "refers to the process in which men [and others], not as recipients, but as knowing subjects achieve a deepening awareness both of the sociocultural reality that shapes their lives and of their capacity to transform that reality" (Freire 1985, p. 93, n2).

19 See details of the outworkings of these groups and conferences in Cooper (2007), Chapter 1.

20 For details of the conversation around each of these see Cooper (2007), Chapter 1.

21 In its simplest form, globalization is "the making global of something." Often the word is used alone without reference to what is being globalized making the term nearly useless in conversation. Properly used, it should be "globalization of." Most commonly globalization refers to the making global of capitalism and the marketplace. When I discuss globalization in this text, I will aim to state globalization of... to specify.

22 The economy is "the system of relations, institutions, and processes concerned with the production, consumption, distribution, and circulation of goods and services to support human life" (Walby 2009, p. 102).

23 Civil society is defined as "a realm of autonomous associations and groups, formed by private citizens and enjoying independence from the government; includes businesses, clubs, families, and so on" (Heywood 1992, p. 311).

References

Althaus-Reid, Marcella. 2000. Liberation Theology. In *The Oxford Companion to Christian Thought*, eds. Adrian Hastings, Alistair Mason, and Hugh Pyper, 387–90. Oxford: Oxford University Press.

Bediako, Kwame. 2004. *Jesus and the Gospel in Africa: History and Experience*. Maryknoll: Orbis Books.

Chambers, Robert. 2017. *Can We Know Better? Reflections for Development*. Bourton on Dunsmore: Practical Action. doi:10.3362/9781780449449

Consultation on Theology and Development. 1970. *In Search of a Theology of Development: Papers from a Consultation Held by Sodepax in Cartigny Switzerland, November 1969.* Geneva: WCC.

Cooper, Thia. 2007. *Controversies in Political Theology: Development or Liberation?* London: SCM Press.

Farmer, Paul. 2013. Health, Healing, and Social Justice: Insights from Liberation Theology. In *In the Company of the Poor: Conversations with Dr. Paul Farmer and Fr. Gustavo Gutiérrez,* eds. Michael Griffin and Jennie Weiss Block, 35–70. Maryknoll: Orbis Books.

Freire, Paulo. 1985. *The Politics of Education: Culture, Power and Liberation,* trans. Donaldo Macedo. London: MacMillan.

Gutiérrez, Gustavo. 2013. The Option for the Poor Arises from Faith in Christ. In *In the Company of the Poor: Conversations with Dr. Paul Farmer and Fr. Gustavo Gutiérrez,* eds. Michael Griffin and Jennie Weiss Block, 147–59. Maryknoll: Orbis Books. doi:10.1177/004056390907000205

Heywood, Andrew. 1992. *Political Ideologies: An Introduction.* London: Macmillan.

Hoffstaedter, Gerhard and David Tittensor. 2013. Religion and Development: Prospects and Pitfalls of Faith-based Organizations. In *Handbook of Research on Development and Religion,* ed. Matthew Clarke, 402–12. Northampton: Edward Elgar. doi:10.4337/9780857933577.00031

Ilo, Stan Chu. 2011. *The Church and Development in Africa: Aid and Development from the Perspective of Catholic Social Ethics.* Eugene: Pickwick.

Kee, Alistair, ed. 1974. *A Reader in Political Theology.* London: SCM Press.

Loffler, Paul. 1974. The Sources of a Christian Theology of Development. In *A Reader in Political Theology,* ed. Alistair Kee, 70–9. London: SCM Press.

Loy, Catherine. 2017. *Development Beyond the Secular: Theological Approaches to Inequality.* London: SCM.

Kothari, Uma. 2005. From Colonial Administration to Development Studies: A Post-Colonial Critique of the History of Development Studies. In *A Radical History of Development Studies: Individuals, Institutions, and Ideologies,* ed. Uma Kothari, 47–66. London: Zed Books.

Mander, Harsh. 2015. *Looking Away: Inequality, Prejudice and Indifference in New India.* New Dehli: Speaking Tiger.

Mitchell, Bob. 2017. *Faith-based Development: How Christian Organizations Can Make a Difference.* Maryknoll: Orbis.

Mombo, Esther. 2009. Religion and Materiality: The Case of Poverty Alleviation. In *Religion and Poverty: Pan-African Perspectives,* ed. Peter Paris, 213–27. Durham: Duke University Press. doi:10.1215/9780822392309-011

Ngo, May. 2018. *Between Humanitarianism and Evangelism in Faith-based Organisations: A Case from the African Migration Route.* London: Routledge. doi:10.4324/9781315561479

Nissiotis, Nikos. 1971. Introduction to a Christological Phenomenology of Development. In *Technology and Social Justice,* ed. Ronals Preston, 146–60. Valley Forge: Judson Press.

Occhipinti, Laurie. 2015. Faith-Based Organizations and Development. In *The Routledge Handbook of Religions and Global Development*, ed. Emma Tomalin, 331–45. London: Routledge.

Paul VI, Pope. 1965. *Gaudium et Spes*. www.vatican.va/archive/hist_councils/ii_vatican_council/documents/vat-ii_const_19651207_gaudium-et-spes_en.html. Accessed September 20, 2018.

———. 1967. *On the Development of Peoples*. Vatican City: Vatican Polyglot Press.

Pew Research Center's Forum on Religion & Public Life. 2011. *Global Christianity: A Report on the Size and Distribution of the World's Christian Population*. www.pewforum.org/2011/12/19/global-christianity-regions/

Rist, Gilbert. 1997. *The History of Development: From Western Origins to Global Faith*, trans. Patrick Camiller. London: Zed Books.

Tomalin, Emma. 2013. *Religions and Development*. London: Routledge. doi:10.4324/9780203831175

Ver Beek, Kurt. 2000. Spirituality: A Development Taboo, *Development in Practice*, vol. 10:1, 31–43. doi:10.1080/09614520052484

Walby, Sylvia. 2009. *Globalization & Inequalities: Complexity and Contested Modernities*. London: Sage.

White, Sarah and Romy Tiongco. 1997. *Doing Theology and Development: Meeting the Challenge of Poverty*. Windows on Theology. Edinburgh: St Andrews Press.

World Conference on Church and Society, Geneva, July 12–26, 1966. *Official Report, with a Description of the Conference. Christians in the Technical and Social Revolutions of Our Time*. Geneva: WCC.

Part 1

Theology

This first section articulates a theology of development.

1 Empowerment

One of the most important pieces to consider within theology and development is the use of power. Power is not a neutral concept. It infuses daily life and can be used for good or ill: to dominate or to empower. Power is enacted in relation with others. Within Christian theology, many people consider God to be all-powerful and all good. Then, the question arises, why does evil exist? Why doesn't God use God's power to prevent evil? This question leads people to wonder whether God exists, whether God is good, and whether God is powerful. Instead, I argue that because God is good, God does not use power to dominate; rather, God empowers creation to act. This chapter argues, from examples of God, Jesus and the Holy Spirit, that a Christian use of power is empowerment: power to, power with, and power within, not power over. Such use of power will shift Christian development practice, as later chapters explore.

"Power over"

Unfortunately, we have often focused on domination by a few over the many, even within Christianity. As Mark Lewis Taylor, U.S professor of theology and culture, has noted, "The world is heavy, then, with social practices that generate and organize death and dying" (Taylor 2011, p. 7). We participate in structures that marginalize and kill. Sometimes we want to move the blame for horrific acts away from humans onto God. However, humans are responsible for their own uses and abuses of power. Christianity and Christians, however, can work toward empowerment rather than domination, following the example of Jesus.

Most people understand power as "power over," in terms of strength and control. In this view, when one exercises power it has a negative effect on another. People tend to see power as "zero-sum;" if I am powerful you are not, and vice versa. "Traditional psychological and sociological concepts of power define it as societal, for example, based on resources, wealth, influence, control, and physical strength (Miller & Cummins, 1992) [...] domination or *power over*" (Norsworthy, McLaren, and Waterfield, 2012, p. 62). If a person has power, s/he will use it to control others, bending others to one's own will. Power is used for an individual's benefit in this definition. Hence, we try to accumulate power for ourselves and we fear losing power to others.

Most of the structures of power we see, as well as the ways in which we interact with others, are in terms of power over. The "nuclear" family, for example, with a husband, wife, and children traditionally understood the husband to be the "head of the household." Religions, including Christianity, developed hierarchical structures where the rules of what to believe or practice come from the top down. Our educational systems have enacted a dominating form of power where teachers hold the knowledge and the students' role is to memorize that knowledge. Paulo Freire, a Brazilian philosopher of education, has countered this understanding, as I note in a further chapter. The state holds power over people, enforcing laws. The state dictates my rights and responsibilities. Pierre Bourdieu, French sociologist, refers to these concepts as the four guardians of symbolic capital. Structures tell us what to do and what not to do; structures constrain our choices and they exercise evil through racism, sexism, and classism, among other isms. We live within these structures of power, tending to see the structures as normal.

This oppressive form of power is more than just economic and political. As Taylor notes from Bourdieu, as humans, we want others to recognize our humanity. With an understanding of power as dominance and zero-sum, we try to accumulate power so that others will recognize us. The key for the dominant group is to set up oppositions with certain values (good and bad) and make the values seem natural, so natural that those people who are labeled negatively may absorb those labels in some sense, called (mis)recognition. If I am recognized as powerful, you must lack power. So I define you in ways that disempower you. You can either accept this definition or be excluded altogether. Bourdieu "speaks of the ratification of domination... as a coercion, one that is 'set up only through the consent the dominated cannot fail to give'" (Taylor 2011, pp. 92–4). It is hard to see and even harder to resist this power. We want to be recognized as human. Sometimes our humanity is only recognized in negative ways. To be dominated is at least to be included, even if in a negative way. We may absorb these negative aspects, assuming them to be part of who we are in the same way we absorb positive reinforcements. Rather than questioning the oppositions, we tend to work within the value system, particularly if the system privileges us.

Because power is key to relations, theology and ethics should not be able to avoid analyzing power. Yet, they often do. Miguel de la Torre, Latino social ethicist, speaking of "ethics from the margins," argues that "power is used to *normalize* what the dominant culture determines to be ethical" (de la Torre 2004, p. 32). Our ethics contain notions of good and bad that include some and exclude others. This normalization has included racism, classism, sexism, and other isms, supporting structures of dominance. Even privileged Christians see this type of power to be normal and supported by Christianity, protecting their own privilege. Using Michel Foucault's (French philosopher) notion of the insane asylum, de la Torre continues "Like the patients in the asylum, the marginalized suffer from their own "madness" – their refusal to conform..." (de la Torre 2004, p. 33) Those who are marginalized are seen as "mad" for not accepting and participating in the system. If I argue against the dominance of men or the church hierarchy or another powerful system, I am seen as the one with the problem. The system is fine. It is also difficult, sometimes impossible, to step outside the system. Some people claim neutrality; however, that claim simply supports the status quo.

We are so enmeshed with the dominating sense of power that we support it in our everyday actions. Foucault, for example, analyzed how this power is not solely from the top-down but is networked, through and within us. To upset this notion of power would upset our systems and our daily lives. We see and participate in a harmful use of power daily, for example, with the killing of black people by police officers in far greater percentages than people of other races or with the senseless death of millions from preventable and curable disease. While some sense a problem with the system, others simply assume it is the way things are in order to "protect" and progress society. In this view, some people have to be excluded for others to survive.

Dealing with power is not as simple as my deciding to behave differently, once I recognize a problem. Danny Burns and Stuart Worsley, U.K. researchers in development, discuss power as a systemic property. "The laws, rules, norms, customs, identities, standards, and so on are elements of a system dynamic which become crystallized like well-worn paths through a forest" (Burns and Worsley 2015, p. 153). We grow up and are educated within these power structures. Without obvious competing narratives or alternatives, we cannot see a different path, even when we feel something is wrong. We can see this tendency even in forms of resistance. I've heard the comment, "the world would be better if women were in charge," and this comment may be true. However, sometimes we don't question the hierarchy itself, we just advocate for change at the top of the pile. It is difficult to see true alternatives.

We not only accept this dominating form of power, we also accept the violence that comes with it. Violence can be both an exercise of power and a reinforcement of power. The violence caused by this domination is seen to be normal. The powerful use violence deliberately as a form of control. Further, the dominant powers in society tend to define violence on their own terms and try to make other forms of violence invisible. In this way, structures determine what violence is, ignoring some forms while exaggerating others. Hence, starving to death from lack of food is not seen as violence, while the violence of stealing food is condemned and punished.

A spiral of violence exists, as Brazilian Archbishop Helder Camara called it (1971). First, daily life is violent, where people die from hunger, thirst, and disease. Taylor speaks of this violence in terms of weight: "the suffering known by the most acute and direct victims of social constraints and oppressive structures" (Taylor 2011, p. 38).[1] The excluded feel the weight of the world bearing down on them. Those of us with privilege can also feel this weight in terms of empathy. Whether or not we do depends on who we see as our neighbor, who we see as Jesus, others with privilege or those excluded from society altogether, condemned to death, if not yet dead. The third chapter explores this concept further.

Those of us who feel this weight also can enact a counter-weight, all of us pressed up against each other, a buttressing weight, according to Jean Luc Nancy, French philosopher. In Camara's terms, a second understanding of violence is the resistance of the people against the first level of violence. Resistance is what tends to be categorized as the first level of "violence" within our society, not the violence of daily life. Resisting domination is condemned. "The word *terror* is usually reserved for those military guerrilla groups fighting the empire... while the empire utilizes the most violent means to suppress the native people" (Raheb 2013, p. 60). The powerful enact their own violence, while publicizing resistance as terror. The label of terror shifts depending on who the powerful understand to be on their side. For example, the USA supported the Taliban in Afghanistan during the Cold War but later shifted to view the Taliban as the enemy. In the Israeli-Palestinian conflict, Israel bombing the Gaza Strip is not considered terrorism, while a Palestinian bomb in Israel would be terrorism. Whether one is labeled a terrorist tends to be determined by whether one supports or resists the dominant system.

The reaction to the resistance, the third aspect of violence, is repression of that resistance by those in power. Because the first aspect of the spiral of violence is not acknowledged by the powerful, and the resistance is seen as the original violence, then this third level of violence is determined to be necessary to maintain control. The violence of everyday existence is ignored because it is an inherent part of the way our systems work. Instead, in the public realm, we tend to argue over the violence we see in the resistance and the repression. Everyday violence remains hidden.

Other countries' experiences of development have often occurred under this dominating form of power by the USA and Europe. The USA, for example, has deliberately used "power over" to dominate in the international realm and it has encouraged other governments to use a dominant form of power to repress their citizens, particularly throughout the second half of the twentieth century. U.S. peace studies professor, Jack Nelson-Pallmeyer outlines five stages of U.S. policy, each using a dominating form of power: from 1946 to 1979, the USA supported dictatorships throughout Latin America, Asia, and Africa. From 1980 to 1991, it continued this support and also enforced its economic policies through the IMF and the World Bank. From 1992 to 1997, the USA mainly used its dominant economic power. From 1998 to 2001, it used this economic power and began to use military power again as well. And in 2001, with the start of the Bush administration it returned to military power, viewing the solely economic push of the Clinton era to be wrong (Nelson-Pallmeyer 2005, pp. 84–98). Both military and economic power have continued to be exerted as dominance in the twenty-first century.

Latin American liberation theology along with other liberation theologies emerged out of the experience of being dominated. The USA directly worked against liberationist Christians in Latin America, according to a 1980 Committee of Santa Fe report: "U.S. foreign policy must begin to counter... liberation theology as it is utilized in Latin America by the 'liberation theology' clergy..." (Nelson-Pallmeyer 2005, p. 75). Nelson-Pallmeyer notes that during the 1980s, people trained by the U.S. Army School of the Americas (SOA) murdered several priests, nuns, and laypeople, including El Salvadorian Archbishop Oscar Romero. "In 2001 the school's official Web page boasted that although many of its critics 'supported Marxism- Liberation Theology-in Latin America" it had been *"defeated with the assistance of the U.S. Army"'* (Nelson-Pallmeyer 2005, p. 75). Liberation theology actually continues to exist and expand around the globe but for a time it disappeared from public view, ignored by the media and academia. Similar experiences of U.S. (and other countries') repressive power occurred across the world and continue today. As noted in the introduction to this book, Ngo argues that this domination is cultural, economic, and political with the North defining "good behavior," encouraging others to behave in the same way, and punishing those who resist. People are expected to submit to the dominant economic, political, and cultural systems.

This dominating form of power has often been expressed through empire. Empire expands beyond the political realm to encompass the economic and cultural realms as well. With empire, it is difficult for citizenship to hold any meaning for many people. Political and economic power combine to exclude the majority of people around the globe. In recent history, "the citizen has been ... emptied of real human content. ... These rights have been absorbed by the 'market'..." (Míguez, Rieger, and Sung 2009, p. 9) When the market is prioritized, as in the capitalist system, citizenship becomes less powerful. One can only exert power with money in the market. "Whoever cannot participate in the market becomes a non-subject, a non-person" (Míguez, Rieger, and Sung 2009, p. 8). If you lack money, you are excluded from the market, and from society, since the market is central to society. In contrast, things can have the rights of persons, in particular, corporations. While capital, especially finance, moves quickly around the world, human beings cannot move freely around the world. People are excluded from or dominated by empire, whether that empire focuses on political control or economic control.

Empires exert power as control. Today, structures control migration, who can go where and when and for how long. Structures also control where we can work and where we can visit. Empires also take control over local resources for the benefit of the powerful. Mitri Raheb, Palestinian Lutheran theologian, cites water rights in Palestine as an example (Raheb 2013, p. 57). Water, along with other goods necessary for life, are not freely available to human beings. Neither are human beings free to move toward a source of water or access land for food. More attention is paid to ensuring money and goods can traverse the globe than enabling people to move.

The exception is when empires encourage their own citizens to expand into the "new" areas, building "new" communities. Corporations use the same logic of control, purchasing subsidiary companies in other places, moving their own employees to a new location, using the natural resources of the new location for the company's benefit.

To expand one's power into a new area, it is easiest to introduce people to that area who already benefit from the system.

According to Raheb, empires have to justify what they do and they do so with ideology and theology (Raheb 2013, p. 64). In our time, this divine purpose is economic: global capitalism and political: good governance. I'll explore both of these concepts in the second half of the book. It is hard to imagine a different use of power. While some theologies argue against dominance, parts of most religious traditions support forms of domination.

Rosemary Ruether, U.S. Catholic feminist theologian, argues that the global North needs to let go of this dominating form of power. To do so, people need "to become critical mediators who press the powerful in their society to let go" (Ruether 2013, p. 44). We need to understand how power as domination works, how it harms, and then work to resist it. One impact of this change in power is that development must be more than about improving the lives of the poor. It must attend to the misuse of power as well. So how does Christian theology help us understand power? To answer this question, we need to return to the Gospels, seeing how Jesus and the earliest Christians lived. Then we need to understand how the radical shift to supporting dominance occurred.

Christianity and power

Judaism, Christianity, and Islam all emerged as resistance movements, in the context of dominating powers. Raheb argues, "It is, in fact, this context of ongoing oppression ...that brought about the birth of both Judaism and Christianity, and across the sea, Islam" (Raheb 2013, p. 86). These major religions, including Christianity, emerged out of oppressive contexts, finding ways to resist. Each religion has also developed strands that have been co-opted to support dominant power. Can Christianity once again help to move toward empowerment or has it crystalized within power as domination?

If we begin with language used to describe God, within Jewish and Christian scriptures, we can start to understand a better notion of power.

> God in biblical tradition is ... [a] source of just relations. The first revelation of God's name happens in the context of the cries of the Hebrew slaves... Moses asks the voice: "What is your name?" The voice answers: "I am Yahweh." ... "I am with you" ... God is a voice in the wilderness of oppression, assuring the downtrodden of God's solidarity.
>
> (Duchrow and Hinkelammert 2012, p. 195)

Within the Jewish tradition, God was with the Israelites in their struggle against slavery. Ulrich Duchrow and Franz Hinkelammert, German liberation theologians, add that even scripture is translated to shift this understanding. Yahweh came to be defined as "I am who I am," to coincide with later Greek philosophy (195). In contrast, to the later misinterpretation, Duchrow and Hinkelammert argue that this relational understanding of God expands when Christianity emerges. Christianity introduces the notion of the Trinity: God, Jesus, and the Holy Spirit. Each part of the divine is unique and together become God. Each part of the Trinity can help us to understand a good use of power, as well as how we are in relation with others and what those relations should look like, as we articulate in the third chapter. No one part of the Trinity dominates the other.

From the beginnings of Judaism, Christianity, and Islam, God was found within the experience of those suffering. For example, Raheb argues, "The revelation the people of Palestine received was the ability to spot God where no one else was able to see him" (Raheb 2013, p. 87). Those with power assumed the divine was on their side, since they were powerful. But the Palestinians found God with them in their struggle. God was not a dominating power. Instead, God walked alongside them and struggled with them, particularly in the person of Jesus. "Jesus revealed this God on the cross, … when he was brutally crushed by the empire and hung like a rebellious freedom fighter" (Raheb 2013, p. 87). Yet, Jesus' death did not end the narrative. Jesus rose again. The powerful killed Jesus but they could not keep him dead. God remained with the people through the resurrection of Jesus, showing the death of death. Jesus' resurrection "ensured that empires were incapable of celebrating their victories, because while they crushed the people they occupied, they weren't able to crush their spirit" (Raheb 2013, pp. 87–8). The empire was unable to kill God. God did not respond by dominating humans; instead, God enacted a different form of power. Jesus did die and Jesus did return. It was possible to resist the dominant powers, by understanding power differently.

Jesus' life and the earliest Christian communities also contradict the notion of power as domination. Jesus taught one could use power differently, empowering others. He modeled this new understanding.

> Jesus's way was one of engagement and involvement through a new way of overcoming, arising from a unique concept of power – the power of forgiveness over retaliation, of suffering over violence, of love over hostility, of humble service over domination.
>
> (Bediako 2004, p. 104)

I elaborate on Jesus' actions in the second and third chapters in relation to justice and accompanying the marginalized. Jesus struggled alongside the marginalized against the dominant forms of power. Jesus' life offers a model of working toward the new heaven and new earth, a model of empowerment, resistance, and love.

Our question is: why do we have this mistaken assumption within Christianity of a dominating form of power being good given the life and death of Jesus? After Constantine, the Roman Emperor converted in the 330s CE, the understanding of Christianity changed from resisting the Roman Empire to supporting it. The basis of Christianity was turned on its head. As Joerg Rieger, a German economic theologian, argues,

> Classical theism envisioned God not only as all-powerful but also as immutable and impassible, qualities that were designed to affirm unilateral and top-down power. ... If Jesus is of the same substance as God, as Constantine proposed to the Nicene Council ... Jesus can now also be envisioned in these terms.
>
> (Rieger 2010, p. 9)

With the concept of consubstantiation, God could have been conceived of as having Jesus' qualities, vice versa, or a mixture of both. However, Jesus became associated with God's dominance, rather than God becoming associated with Jesus' resistance to domination. With Emperor Constantine's influence, the hierarchy of the church solidified, the theological options narrowed and the chosen option supported the powerful. Thus, all the examples of Jesus' resistance of power and siding with the marginalized can be set aside in favor of the new model. God is found on the side of the powerful, not the powerless. God exerts power over; God does not empower (except for the emperor). We even celebrate Jesus birthday on December 25th, the date originally set aside to celebrate the Roman war God, Mithras. From the leader of a resistance movement, Jesus became a war God. Understandings of God and Jesus were inverted to support imperial power. The cross and the sword could now proceed together. There were now two strands within the Christian tradition: one of dominance and one of resistance.

Let's explore the strand of resistance, which has continued to exist throughout Christian history, challenging the dominant theologies. God did not simply create and then sit back to watch what would happen; God empowers humans to act. Some theologians understand this concept as the "immanent God": "God dwelling within creation"

(Moe-Lobeda 2002, p. 15). God continues to work within history through human beings. God is present now. God is actively involved in this world empowering humans to work against injustice through the Holy Spirit. God walks alongside us daily, as we make our choices for good or evil.

Second, God is on the side of the marginalized, as the third chapter will explore. "The God whom we know in the bible is ... a God who intervenes in history in order to break down the structures of injustice" (Gutierrez, 1974, p. 116, quoting from a Manifesto of the Bolivian Methodist Church). God empowers and encourages the "least of these." God also suffers with the least of these. Dietrich Bonhoeffer, a German Lutheran theologian, killed by the Nazis for his participation in a plot to assassinate Hitler, also stated that God is found on the "underside," which is where theology should be too. Because God loves all human beings equally, God struggles alongside people who are treated least like humans. It is not that God loves the poor more than the rich; it is that God wants each of us to be able to live life abundantly. And so God walks closely with those who are excluded from this abundant life. Further, those of us with privilege move away from God in our actions.

Most theologians agree that sin separates humans from God. However, the question of what moves us away from God is contested. One strand narrowed to a notion of individual sinful actions and individual salvation. The other strand remained broad. Sin can be defined as injustice in all its forms, when humans choose domination over empowerment. "Sin is regarded as a social, historical fact, the absence of brotherhood[2] and love in relationships among men, the breach of friendship with God and with other men" (Gutierrez 1974, p. 175). Rather than choosing to love others, we prioritize ourselves. Humans choose unjust relations, sinning. Part of this sin is ensuring life abundant only for some. To become closer to others and to God we need to choose justice. Humans come to know God through knowing each other. Thus, when we have unjust relations with other we are separating ourselves from knowing God. "In our tradition, the essence of sin is in its being an antisocial act, ...injury to the interests of another person. ... But human beings are... created in God's image, so social sin is also sin against God" (Bediako 2004, p. 26). Each person we encounter is Jesus, as we explore further in the third chapter. Our relations with others determine our relation with God. For each of us to live life abundantly, we need to craft just relationships. In order for us to build a good relationship with God, we need to build good relations with others.

Because structures are created and sustained by humans, sin exists in structures as well as in human action. As Séverine Deneulin, a French professor of development, articulates, it is impossible to avoid structures of sin. We live within these structures, which constrain our choices. "Overcoming structural sin requires a collective turning away from sin and coming back to God" (Deneulin 2013, pp. 62–3). I, as one individual, cannot overturn structures of sin but together we can work toward a new heaven and a new earth. We can resist structures of sin, such as an economic system focused on profit rather than people; we can work to change the structures together.

Injustice stems from the inequality of power relations and the misuse of power between humans. In God's new heaven and new earth, humans will live without sin. Each human being will be empowered. This concept is justice. For Nancy, "justice is thus the return to each existent its due according to its unique creation, singular in its coexistence with all other creations" (qtd in Taylor 2011, p. 50). Each of us should be able to live life abundantly. We explore charity and justice in this context in the following chapter. God created all to live together. No one person needs to hold power over others. Each can have his/her due without harming another; this justice will be fully enacted in the new heaven and new earth. The Christian task is to work toward justice now.

Where some traditional theology has emphasized the afterlife, other theology focuses on also working toward God's new heaven and new earth now. Humans are called by God to work toward justice, rather than sit back and wait for God to rescue the world. Humans will not bring about the new heaven and new earth; God will. However, humans are responsible for pointing toward this new heaven and new earth where justice will rule. Development can be considered in this context, a working toward enabling each human being to live life abundantly.

Salvation, in terms of repentance from sin and conversion to a new life, then, can be conceptualized in terms of the community, rather than solely one individual. "Salvation, in the Hebrew Bible, is understood as *shalom:* ... peace ... embraces all the levels of liberation – good health, security, good relations, prosperity, contentedness" (Gorringe 2004, p. 154). Salvation is holistic; it is not solely spiritual leaving injustice intact. The third chapter explores the concept of conversion as requiring action. The fifth chapter will refer to aspects of class in arguing that economics needs to be framed around enabling each human to live life abundantly. While few people argue that humans should be separated by class, many accept the current global economic system, which increases the disparity between classes. The sixth chapter will refer to other areas of disparity of power, arguing for attention to the intersections of injustice.

Humans move closer to or further away from God based on their relations with each other. Jesus died and was resurrected to show that salvation from sinful structures and actions is possible. God overcame sin, not through a dominating use of power, but by becoming fully human in the form of Jesus and modelling resistance to evil. For humans to overcome separation from God, we need to work toward this new heaven and new earth, in our communities and our structures.

The Holy Spirit is our guide in this work today; God remains present in this world. We are not attempting these changes alone. As Grace Ji-Sun Kim, a Korean-American theologian, explains, "There is power within the Spirit which can lead to greater accomplishments and transformations within us and in society, nature, and world" (Kim 2013, p. 71). The Holy Spirit can work within us to guide our actions. The book of Acts shows how the Holy Spirit continues to move in this world, empowering people after Jesus' death and resurrection. The Apostle Paul's letters confirm that the Holy Spirit is with the new believers. "The Spirit works to give illumination and divine revelation in the face of affliction (1 Thess. 1:6; 1 Cor. 2:10–12; 2 Cor. 3:14–17)" (Kim 2013, pp. 72–3). The Holy Spirit empowers humans to work toward the new heaven and the new earth, resisting dominant forms of power, even those dominant forms that have developed within the Christian tradition. Listening to the Spirit guides us toward the new heaven and new earth. Ignoring the Spirit perpetuates injustice.

God is in the world, amidst the suffering and loss. God is a victim with the victims. The third chapter explores this concept further. God is within each of us. As we treat others, we treat God. One of our responsibilities is to rescue God from suffering by valuing each human life, simply for one's humanity. "If we find an individual who is a poor, black, lesbian, AIDS-infected, disabled, ugly, and old prostitute, and still we see this individual as a human being in her fundamental dignity, we will be undergoing a spiritual experience of grace... and faith" (Sung 2005, p. 5). The adjectives used to describe this human being used by society to exclude and marginalize some humans; however, this person is created in the image of God and is fully human. We need to include each human being and work against domination. In working to include and respect human beings, we are including and respecting God. Our current system tends to value more highly those people who benefit from the economy. Yet each human being is of equal value, simply because s/he is created in the image of God. One's value as a human being is not changed by any particular contribution or bad behavior.

Taylor (2011) adds a term to Nancy's notion of weight: shifting weight. One could imagine a stone arch. If just one block comes askew, the weight could tumble. Building on this concept, I want us to think about using this notion of shifting weight deliberately within development. Rethinking how we should use power could help the oppressive structures tumble. Then the question would be: how do we rebuild?

So what does this understanding of God, Jesus and the Holy Spirit mean for a definition of power?

Empowerment

Within the weight of the world, the privileged feel they are entitled to certain spacing, a spacing that the oppressed lack. The privileged ensure they have this spacing. Hence, the oppressed feel "concentration" (Taylor 2011, p. 41). The oppressed and marginalized prevented from having the space in which to live a full life because that space is taken up by those with privilege. For Nancy, this situation is evil: "absolute evil, a wound opened up on itself" (Nancy qtd in Taylor 2011, p. 41). This weight is unjust, the injustice perpetuated by the world. Living, being fully human, requires space to be human. Evil is the crushing, shrinking of this space, accomplished through unjust networks of power. We need to expand the space so each human being can be fully human. This spacing also requires those of us with privilege to understand "enough," as the next chapter explores.

The oppressed and excluded can weigh in by struggling against the oppression. Taylor, 2011, gives the example of a prisoner hurling a food tray (Taylor 2011, pp. 35–7). This weighing in is amplified as Taylor writes about it, bringing it to a wider audience. Bringing a voice to the voiceless, amplifying that voice is one way to begin a good use of power. One role for development agencies is to amplify the voice of the poor. For Taylor, art is another piece of weighing in, telling true stories, imagining an alternate world. We need to listen to those without power, as they imagine a good world; development can begin here.

In contrast to the common definition of power above, power can be enabling and shared. A good use of power enables people to live a full life. If God is good, God will use power for good, empowering humans. "Feminists typically redefine power as *empowerment* or *power to* (Yoder & Kahn, 1992). From this view, power is the ability or capacity to do things" (Norsworthy, McLaren, and Waterfield, 2012, p. 62). Power is not zero-sum. It is possible for each of us to be able to act.

I distinguish three aspects of empowerment. The first is for a person or community to realize on their own that they lack power. The second is moving forward in the process of understanding the surrounding world and society. The third is action to change one's place in the surrounding world.

Further, this type of power encourages sharing. Power should not be enacted as domination or control, one individual crushing another. We can empower each other. In this model, power increases mutually as we work together. We become more able to make decisions, to live together and thrive together. We can each achieve the capacity to be fully human.

Empowerment includes power to, power with, and power within. Power to is the ability to do something, including making decisions. Power within "is a process by which dominated people shake off the control of others ..., laying hold of their own innate power and goodness" (Kim 2013, p. 7). We inherently have capabilities, which we and others should recognize. Finally, power with "is the development of ways to share power that ... can mutually affirm one another" (Kim 2013, p. 71). There are many forms of power as empowerment rather than domination. Each person can have the ability to act, recognize their own personhood, and work together. None of these understandings of power include domination.

One important piece of empowerment is that it recognizes all human beings as human and respects each human being. Recognition is when "a human being acknowledges the existence and presence of other human beings" (Hankela 2014, p. 356). First, we need to acknowledge that each human being is actually human and of equal value simply by being human. Instead of setting up ethical systems that privilege some and exclude others due to race, class, sex, or other intersections, we can acknowledge human beings as equally human. Second, "if your presence as a human being is truly recognized, your humanity ought also to be respected through respecting your needs, moral/religious values and abilities" (Hankela 2014, p. 357). Recognizing another as human should lead to respecting the other's humanity, not simply assuming one's own perspective is correct or should be dominant. We need to respect each other as human beings through our actions. Recognition and respect are the starting points for living and working together in community.

Returning to language about God, all-powerful means one has complete power. All-powerful can mean the ultimate power to empower others. God may be all-powerful: empowering, working through history to enable humans to do good. God would not be good if God used God's power to dominate others and bend them to God's will. Jesus provides examples of this good use of power throughout the Christian scriptures. First, Jesus heals, feeds, and associates with the poor and marginalized, empowering them. Second, Jesus dies at the hands of the powerful and returns, resurrected. Jesus does not use divine power to force the authorities to bend to his will. Rather, Jesus shifts the narrative and the use of power by returning. Jesus overcomes even death.

If we take the Christian church as an example, we can see that it has evolved to choose a dominant hierarchical power structure rather than an empowering one. Timothy Gorringe, Anglican priest and theologian, states that the New Testament had a different way of being church, where each member of the community was equally important. As Paul stated,

> Power is vested in "the body of Christ".... God has so arranged the body, giving the greater honour to the inferior member, that there may be no dissension within the body.... If one member suffers, all suffer together with it (1 Cor. 12:24–26).
>
> (Gorringe 2004, p. 152)

In Christian community, we are each equal as human beings made in the image of God. Together, we form the body of Christ. No one of us is more or less important than another. Early on, the groupings of Christians were not to be hierarchical. Each person had a gift to offer the community, none more important than another. As Christianity expanded, the church became more hierarchical excluding some. It has excluded people from leadership, excluded particular beliefs as "heretical" [wrong], and it has excluded other religious traditions by declaring them to be untrue. In contrast, Christianity can be a model of inclusion and empowerment.

Christian FBDOs can be too. As I explore in later chapters, FBDOs can ensure their work with the poorest is based on empowerment, that their own organizations are based on empowerment, and that they advocate for the ending of power over. This change would shift the practices of FBDOs.

Conclusion

Thinking about "power to," "power within," and "power with" rather than "power over" means there would be a radical shift in the structures and institutions governing our lives, tumbling the archway. The question would become: what can we enable people to do? How can we move from concentration to the spacing we each need to be fully human? We need to think about how to shift toward a good use of power within structures, in our context: development. The second part of this book addresses this task.

The first step is to recognize the imbalance of power in the current system. Freire calls this process conscientization, becoming aware of the wider world and the community's place in it. The second step would be to accompany the marginalized and work toward the new heaven and new earth together. Immersing ourselves into the pain and suffering of the world, we can work to end the evil of injustice. How humans behave in this world is critical in terms of moving us further away from that new heaven and new earth or toward it. God is here inside the world, empowering humans to work toward justice. One example of how to live together can be seen in the celebration of the Eucharist. Each person receives a share of the body and blood of Christ. All who partake of the body and blood are empowered to live out community in God's new earth. Power, in this new earth, can be shared. Power is to choose life over death.

Having considered a good use of power, we turn to how empowerment might affect the concepts of charity and justice.

Notes

1 He derives this concept from John Luc Nancy (2007).
2 Older texts cited here do not use inclusive language. Many of these authors do, however, in newer texts.

References

Bediako, Kwame. 2004. *Jesus and the Gospel in Africa: History and Experience*. Maryknoll: Orbis Books.
Burns, Danny and Stuart Worsley. 2015. *Navigating Complexity in International Development: Facilitating Sustainable Change at Scale*. Bourton on Dusmore: Practical Action. doi:10.3362/9781780448510
De La Torre, Miguel. 2004. *Doing Christian Ethics from the Margins*. Maryknoll: Orbis Books. doi:10.1177/0040571x15578832h
Deneulin, Séverine. 2013. Christianity and International Development. In *Handbook of Research on Development and Religion*, ed. Matthew Clarke, 51–78. Northampton: Edward Elgar. doi:10.4337/9780857933577.00009
Duchrow, Ulrich and Franz Hinkelammert. 2012. *Transcending Greedy Money: Interreligious Solidarity for Just Relations*. New York: Palgrave. doi:10.1057/9781137290021.0019
Gorringe, Timothy. 2004. *Furthering Humanity: A Theology of Culture*. Burlington: Ashgate. doi:10.4324/9781315254807
Gutierrez, Gustavo. 1974. *A Theology of Liberation: History, Politics, and Liberation*, trans. Sister Caridad Inda and John Eagleson. London: SCM Press.

Hankela, Elina. 2014. *Ubuntu, Migration and Ministry: Being Human in a Johannesburg Church*. Boston: Brill. doi:10.1163/18748945-02901019

Kim, Grace Ji-Sun. 2013. *Colonialism, Han, and the Transformative Spirit*. New York: Palgrave Pivot.

Míguez, Néstor, Joerg Rieger, and Jung Mo Sung. 2009. *Beyond the Spirit of Empire*. London: SCM Press.

Moe-Lobeda, Cynthia. 2002. *Healing a Broken World: Globalization and God*. Minneapolis: Fortress Press.

Nancy, Jean Luc. 2007. *The Creation of the World or Globalization*. Albany: State University of New York Press.

Nelson-Pallmeyer, Jack. 2005. *Saving Christianity from Empire*. New York: Continuum.

Norsworthy, K., M. McLaren, and L. Waterfield. 2012. Women's Power in Relationships: A Matter of Social Justice. In *Reproductive Justice: A Global Concern*, ed. J. Chrisler, 57–75. Denver: Praeger.

Raheb, Mitri. 2013. *Faith in the Face of Empire: The Bible through Palestinian Eyes*. Maryknoll: Orbis. doi:10.1111/erev.12123_3

Rieger, Joerg. 2010. *Globalization and Theology*. Nashville: Abingdon Press.

Ruether, Rosemary. 2013. A U.S. Theology of Letting Go. In *The Reemergence of Liberation Theologies: Models for the Twenty-First Century*, ed. Thia Cooper, 43–8. New York: Palgrave. doi:10.1057/9781137311825.0010

Sung, Jung Mo. 2005. The Human Being as Subject. In *Latin American Liberation Theology: The Next Generation*, ed. Ivan Petrella, 1–19. Maryknoll: Orbis Books.

Taylor, Mark Lewis. 2011. *The Theological and the Political: On the Weight of the World*. Minneapolis: Fortress Press. doi:10.2307/j.ctt22nm9rl

2 Justice

Wealthy and powerful people tend to see helping others as an option, something good we could do, rather than understanding inequality as injustice. Having defined a good use of power as empowerment, we now consider what empowerment means for charity and justice. We will begin by defining poverty, wealth, charity, and justice in the biblical and early church contexts and compare that to the common definitions now. Then we can assess what we can shift to achieve a form of economic development supported by Christian theology. What does Christian theology have to say about our current global situation where the world's ten wealthiest people have more money than half the world's population? (Jacobs 2018).

Biblical understandings of charity and justice

The biblical text defines poverty expansively. The Hebrew terms for poverty include notions of "humility, neediness, lowliness, thinness, and dependency" (Van Til 2010, p. 63). It is more complicated than lacking resources. Poverty includes lacking dignity and freedom. Noam Zohar, Israeli Jewish professor of ethics, states that it means ""suffering" or "misery," ... such as childlessness or oppression" (Zohar 2010, p. 205). He notes that the King James Version uses the term "affliction" (Zohar 2010, p. 205). Thus, the poverty of some is caused by others, similar to using a dominant form of power over others. Poverty should be fought against, which requires fighting against domination. The early rabbis continue this notion of poverty including: "(1) unsatisfied needs/yearnings, (2) marginalization and humiliation, or (3) lack of property" (Zohar 2010, p. 209). Poverty was both material and social. Alleviating poverty, in the Jewish tradition, is a commandment from God; it is not optional. Poverty should not be accepted as part of the system.

In the New Testament, this expansive definition continues. Poverty is the inability to live life abundantly. For example, in Luke, poverty is isolation, being a leper or deaf is included (Van Til 2010, p. 63). Anything that causes a person to be excluded is an aspect of poverty. This understanding of poverty then links to the notion of wealth. Wealth, properly understood, is the ability for each person to live life abundantly, rather than individual accumulation. For example, property was to be held in common, not owned individually. Any land was to benefit the community.

Jesus was clear that we need to reduce both riches and poverty. Phillip Goodchild, a U.K. religion professor, cites several examples from the Gospels: "Woes were proclaimed to the rich; blessings to the poor (Luke 6.20–26). ...It was impossible for the rich to enter the kingdom of God (Mark 10.25)" (Goodchild 2009, p. 2). Jesus and the Gospels were not shy about condemning poverty and wealth. Jesus sided with the poor and marginalized, critiquing riches, and Jesus worked toward justice. To follow Jesus, one could not keep riches and leave others excluded. The choice was between following God and following money.

The biblical text has a holistic rather than narrow notion of justice. Jose Miranda, a Mexican biblical scholar and liberation theologian, discussed charity and justice in the biblical context. ""To give alms" in the Bible is called "to do justice"... Almsgiving ... was a restitution that someone makes for something that is not his" (Miranda 1977, pp. 14–15). Accumulating riches is unjust. When we "give" to someone who is poor, we are simply returning to them what they are owed. The poor should have the same access to all (economic, political, social...) resources as the rich. It is this imbalance in power and our desire to "keep" power that leads us toward charity, rather than enacting justice.

However, one can only be charitable (giving of one's own goods to another) when there is already a just situation. When there is injustice, one is required to give, it is not a choice because that good already belongs to another. What you have in excess of what you need is not yours.

Horsley notes three important principles of economic justice in the Hebrew Bible. First, the earth belongs to God; humans can make use of it. Second, land was to be distributed among the community, not accumulated by one or a few landowners. Third, if a person needed to borrow, another should lend freely. Charging interest was unjust (Horsley 2009, pp. 38–41). Combined with the notion of Jubilee, whereby land was to be redistributed and debts forgiven, these principles aimed toward justice in community. Justice meant each person should have access to the resources necessary for life; society needed to protect this balance.

The Jewish tradition continued to emphasize how to retrieve economic justice through alleviating poverty. Maimonides (1135–1204), Jewish philosopher, explained how one should go about alleviating poverty from the best option to the worst option.

> The greatest level, higher than all the rest, is to fortify a fellow Jew and give him a gift... or find work for him, ... so that he will not fall and be in need. One level lower than this is one who gives tzedakah to the poor and does not know to whom he gives, and the poor person does not know from whom he receives...
>
> (Maimonides, 10:7–14)

And so on, to the cranky giver. The best way to alleviate poverty is to prevent it in the first place, helping those people most at risk. Unfortunately, we have enabled people to fall into poverty already, so we need to relieve it. Redistribution is key here, without causing indebtedness or dependency. The next chapter analyzes, however, that beyond giving economic resources, one also has to build just relations with others. Poverty is broader than economics.

The earliest Christian communities carried on Jesus' and early Jewish understanding of wealth, poverty, and justice. In Acts, Christians are described as holding goods in common in their community and meeting the needs of the poorest. Justo González, a Latino Methodist theologian, spends time analyzing Acts 2:42 and 43–47:

> The four things listed here, the apostles' teaching, *koinonia*, the breaking of bread, and prayers, are taken up in almost the same order in verses 43 to 47, where we are told ... (2) that all who believed had all things in common.... The *koinonia* is a total sharing that includes the material.
>
> (González 1990, p. 83)

These early communities saw the outworking of the Holy Spirit, shared their possessions, worshipped together and communed together. Each of these four aspects was important to the community: spiritual, material, and social. It was not enough to listen to the Holy Spirit or worship together. It was also important to build community together and meet the needs of others. These communities were distinct from the surrounding culture[1] in the community's care for one another. In Roman culture, property rights were absolute. The dominant culture gave priority to the male citizen of property, and many people were left at the margins.

Over time, these "demands" of the Christian community shifted toward "advice." While the four aspects remained the ideal, as more people became Christian, they also wanted to maintain relations with their families and friends in their own community. "In some instances emphasis shifted from commonality of goods – *koinonia* – to almsgiving. ... The criterion they most often use is that one should keep for oneself only what is necessary" (González 1990, pp. 226–7) While Christian communities no longer held everything in common (a common purse), Christians were still to give away all but what they needed. People did not have to physically leave their old life but their actions would change and their new life would be oriented to meeting the needs of others.

Throughout the early church, as wealthy people began to become Christians, the question of how the rich could be saved became more urgent. The answer was give your wealth to the poor. John Chrysostom, Archbishop of Constantinople (c. 347–407) agued more than 300 years after Jesus that:

> This is also theft, not to share one's possessions. Perhaps this statement seems surprising to you, but ...I shall bring you testimony from the divine Scriptures, ... the failure to share one's own goods with others is theft and swindle and defraudation.
>
> (Chrysostom 1984, p. 49)

For centuries, the understanding that each person should be able to live life abundant remained, as did the notion that Christians were responsible to give away all that they had, apart from what was needed. Limiting wealth and alleviating poverty remained important to the faith.

Further, early theologians argued that lending and borrowing should be free from interest. According to Augustine (354–430), a Roman African theologian, "Lending money on interest is evil. On the contrary, one should give freely; if one cannot give, then one should lend without expecting a profit in return" (González 1990, p. 217). The prohibition on interest, part of the Jewish tradition, continued in the Christian communities. If you are unable to give away what someone else needs, you should at the very least lend it to them without expecting anything in return. And if the person could not return the good, you were not to demand it.

Consistently, property was to be shared, held in common. The surrounding Roman culture understood private ownership of property to be unrestricted; not even to be taxed. In contrast, Christian theologians continued to argue that God owns the land, we have use of it temporarily, and that use is for the common good (González 1990, pp. 227–8). If a Christian owned property, it needed to be used for the good of all, not solely for themselves.

How did we get so far from the biblical understandings? Similar to the definition of power, it is only after Christianity became intertwined with the ruling system with Emperor Constantine's conversion in the fourth century that interpretations of wealth, poverty, justice, and charity changed. God no longer owned the earth; a small subset of humans did. As Christianity became the religion of the elite, it became powerful and supported the powerful.

Throughout Christian history, some Christians have continued to argue against inequality. See, for example, *Radical Christian* Writings (Bradstock and Rowland 2002). Some theologians also continued to emphasize the need to focus on the poor. John Calvin, French Reformation theologian, for example, stated: "Whatever man you meet who needs your aid, you have no reason to refuse to help him. ... The image of God, which recommends him to you is worthy of your giving yourself and all your possessions (Calvin 1960, p. 696). Each human being is made in the image of God and deserves to live life abundant. Our task as Christians is to meet the needs of others. However, these words went unheeded by most Christians with wealth. Many human beings were excluded, some seen as less than human; the powerful even considered some groups of people to be ordained by God to be slaves.

The biblical texts and many theologians argue for justice, which requires alleviating both poverty and wealth.

Common understandings of charity and justice today

Charity, as the word is used today, focuses on changes to the poorest, like poverty reduction, which require help from the wealthy. There tends to be a distinction between the giver and the person being helped in a discussion of charity. The poor need the help of the rich but the rich do not need the poor. The Bible rejects this understanding; instead, it urges justice.

Justice is commonly divided into three types today: commutative, distributive, and social. Commutative justice is holding to a contract: "it involves the right to get you pay for and the obligation of paying for what you get" (Pilarczyk 1997, p. 43). Once there is a contract, it must be upheld by both parties. Distributive justice considers how goods should be distributed in community; "the justice of commonality or of general sharing" (Pilarczyk 1997, p. 47). Distributive justice thinks about basic needs, expanding beyond holding to agreements and considering what resources each human might need to survive. Finally, social justice considers "the right and obligation of individual persons to be involved in determining the way in which larger social, economic, and political institutions of society are organized" (Pilarczyk 1997, p. 52). Beyond our need to survive, we also each need to be able to participate in community life and help organize the structures of life. Commutative justice then means both parties adhere to an agreement. Distributive justice means the good of every person in the community is considered economically. Social justice means the good of every person in the community is considered broadly.

Justice is important to the capitalist system, when narrowly defined as commutative justice. "In labor markets this includes paying people their agreed-upon wages... In product (and service) markets sellers have an obligation to accurately represent products" (Claar and Klay 2007, p. 219). Commutative justice does not demand that I receive a wage that will enable me to survive. Commutative justice simply demands I be paid what the employer agreed to pay. If the price of gas is $3.25 per gallon, then that is what I pay if I buy gas. If I get a mortgage on my house, commutative justice demands that I uphold that contract, despite my ability to pay. Commutative justice does not demand that the bank only charge what I can afford.

A capitalist economic system includes private ownership, a focus on the market, profit and growth, and separation between labor and ownership.[2] After the separation of ownership and work came the corporation. Then that corporation became a person in the USA. Corporations are primarily responsible to produce profit for shareholders.

A different theology has arisen within the capitalist system, articulated by Michael Novak, a U.S. Catholic theologian, in particular, which tends to only leave room for individual acts of giving. First, alongside the strand of the Christian tradition that prioritizes the individual, capitalism also focuses on **the individual**. "In order to create wealth, individuals must be free to be other... when they form communities, they choose them" (Novak 1982, p. 355). The individual exists first and then can choose whether and how to be in community. I only relate to others in the way that "I" choose. One does not work for community or in community but for oneself. In the similar strand of the Christian tradition, the emphasis is on an individual getting to heaven, through individual salvation.

Second, the freedom of the individual is **to participate in the market** without any interference. Capitalists argue that: "markets work well in directing resources to their most highly valued uses" (Claar and Klay 2007, p. 136). The law of supply and demand ensures producers and consumers will come to agreement. Of course, this participation depends on access to money and the availability of the item I desire. Capitalism, some argue, is liberative, because of this freedom from interference.

Third, capitalism prioritizes **rational self-interest** and assumes that individuals "maximize" our self-interest. "Individuals use rational expectations about the future in deciding what economic actions to take" (Claar and Klay 2007, p. 136). That is, I always look out for myself and analyze all decisions to make sure it is in my interest. Further, my self-interest is to make as much money as possible to buy as many goods as possible. Theologically, this ability to prioritize myself is considered a good within capitalism because it means I can live according to my values and no one else's.

Fourth, capitalism encourages "insatiable desire," focused on **unlimited wants** rather than needs (Bell, 102) "Economists use the term wants to refer to human desires generally, rather than trying to distinguish between needs and desires for goods and services" (Claar and Klay 2007, p. 168). The market is not organized around meeting basic needs; it is organized around wants. Humans needs are limited but the market responds to wants rather than needs. Hence, growth is important. More is better. Buying more is better, producing more is better, accumulating more money is better. What helps the system grow is good. As Friedrich Hayek, an Austrian philosopher, argued, "He is led to benefit more people by aiming at the largest gain than he could if he concentrated on the satisfaction of known persons" (Hayek 1978, p. 145). Rather than be troubled by the accumulation of money and power, capitalism sees accumulation as good. This theory argues that creating as much wealth as possible is important because it will trickle-down to generally improve the lives of the poor.

So fifth, **competition** enters. "Competition is the natural play of the free person" (Novak 1982, p. 347). Novak, and others, argue that competition causes individuals to work to be their best selves. To accumulate as much as possible, I need to work harder, and do better, than anyone else. Resources are scarce and if I do not accumulate them someone else will. If I get a job, others do not. If I buy those shoes, others could not buy that pair. All of this accumulation is good within capitalism. I am competing with every other human being, thus the notion of a common good is nonsensical. If I give something up, someone else will take it, buy it, etc. We even see this competition and accumulation in the concepts of human capital or social capital; how can

I use my network of friends, colleagues, and workers, and students, to move myself further forward?

Sixth, according to capitalist economics, God has not created an earth with enough resources for every human being to meet their unlimited desires; hence, resources are scarce. Rather than seeing this assumption as a problem, capitalist theologians see scarcity as a good thing because **scarcity** encourages each human being to be competitive and creative. It is not simply that there are limited resources in the world, which ecologically might be true; it is that my desires are ever growing. I will never have "enough"; therefore logically, there cannot be "enough" to satiate my desires. "Money ... combines the promise of wealth with the threat of poverty. ... For one who holds money and buys, ...once the purchase has taken place, however, the condition is replaced by one of the absence of money" (Goodchild 2009, pp. 110–11). Even when one has money, anxiety remains because as soon as I purchase something the money is gone. No matter how much is accumulated, the fear of poverty remains. No matter how many resources may be available in the world, resources will always be "scarce" for the capitalist system because there is no end to the desire to accumulate.

Seventh, a human being's value can be costed. In fact, concentrating on economic growth means **some humans will be sacrificed**. Some people die from poverty and exclusion, so that others can live. Within capitalism, a human being only has value in the physical or intellectual labor s/he can provide. There, lives are not of infinite value. "Were that the case, then we should immediately outlaw all driving anyway in the United States... Nevertheless, we do not" (Claar and Klay 2007, p. 63). This argument misses the fact that we have enacted laws to improve vehicle safety, for example, seat belt laws. In the USA, the approach to consider public safety in car manufacturing shifted dramatically after Ralph Nader published *Unsafe at any Speed* in 1965. Instead, Victor Claar and Robin Klay, U.S. economics professors, describe how economists calculate the value of a human being's life. "The first approach uses the wage and salary earnings of individuals as a proxy for those individuals' value to society" (Claar and Klay 2007, p. 63). Here, one calculates the value of a human life to be the equivalent of lost salary. Hence, the lives of those who make more money are worth more than the lives of the poor. The second way is an "attempt to evaluate the anticipated benefit of a life-preserving project by assessing how much the relevant individuals value the reduced risk of death..." (Claar and Klay 2007, p. 64). Here, one tries to estimate how much a person would pay to stay alive, which would of course depend on what a person could afford. For people who support capitalism, some human beings will be inevitably harmed.

For economists, the medical principle of "do no harm" does not exist. Instead, the value of a life (in terms of how much a person might earn) is weighed against the cost of protecting and maintaining that life. Capitalist theologians argue that capitalism is the best economic system at controlling human sinfulness. "The world is not going to become – ever- a kingdom of justice and love" (Novak 1982, pp. 341–3). We live in a sinful world; in this view, we simply need to accept this fact. In contrast to theologians who argue that God is working in the here and now and that humans are called to work toward God's new heaven and new earth, Novak rejects that possibility.

Some theologians have correlated Jesus' death on the cross with commutative justice. This theology seems to assume God is subject to an economic order. For these theologians, Jesus died on the cross to pay off the debt of humanity's sin. God had asked humans to obey. Humans sinned by disobeying God, incurring a debt. For humanity's sins to be forgiven someone had to pay the debt: commutative justice.

However, for other theologians, the death of Jesus was not an example of commutative justice. Instead, according to Anselm (1033–1109), a Benedictine monk and Archbishop of Canterbury,

> Christ's work on the cross is a display of the plenitude of divine charity (John 3:16) ... The atonement ... is a matter of God's ceaseless generosity. ... Christ is not our offering to God but God's offering to us (Rom. 5.8).
>
> (Bell 2012, p. 150)

God's justice for humans was not commutative; rather God freely provides justice. It is an expansive notion of justice. God wanted good for humans. God loved humanity so much that Jesus lived to show us how to work toward justice and died to enable us to be fully human. We can now work toward a new heaven and a new earth.

We follow Jesus' example as we continue working toward the new heaven and new earth.

> Salvation was both a spiritual and material gift from God .... The liberation from the net of sin and evil, as well as from the clutches of poverty and suffering caused by injustice was central to the proclamation and ministry of the Lord.
>
> (Ilo 2011, p. 185)

Humans were rescued holistically: spiritually, materially, and socially. This rescue was the good news. Humans now had the ability to live life in all its fullness, if they so chose. Jesus' life is a model for our lives. We need to work to empower others and shift toward justice as we accompany the marginalized. God's new earth became a possibility. Humans can choose to work toward the new heaven and new earth or away from it.

Returning to a Christian conception of charity and justice

Our notion of justice must expand. Understanding justice as more than commutative would turn much of capitalist theology on its head. What would be an alternative theology that does not focus on (1) The individual; (2) Freedom for the market; (3) Self-interest; (4) Unlimited desire; (5) Competition; (6) Scarcity; and (7) Human sacrifice, leading only to individual acts of giving? Here, I highlight communion, freedom for the common good, desiring abundant life with God and others, abundance, and humans as God's good creation.

Communion

No individual can survive alone. We all rely on networks: simple examples include using a road or being born. "Individualism is merely the myth of the powerful; ... their identity is invariably built in relation to others and, more specifically, on the backs of others" (Míguez, Rieger, and Sung 2009, p. 48). In the economy, workers produce profit for owners. Socially, we are born into relations with those who raise us. We also make use of much provided for us, rather than what we produce ourselves: roads, streetlights, electricity. Yes, individuals pay taxes to use public goods but if others did not labor to produce these goods, we could not make use of them. Further, even within capitalism, an individual can only calculate their worth in relation to others, competing with others. Nothing is done "alone."

At the heart of the Christian community is the concept of communion. This sharing of the Eucharist provides a model for how sharing can occur in community today. As the Eucharist shares the body and blood of Christ equally amongst the Christian community, so too could all goods be shared in community. Further, as the Eucharist is a sacred sharing of God's body, so too is the sharing in community sacred. According to Martin Luther, German Reformation theologian, "The sacrament has no blessing and significance unless love grows daily and so changes a person that he is made one with the others" (Luther 1989, p. 251). Christians celebrate communion to remember the good news that Christ's life and death brought. However, it is not enough to remember it solely in that moment; it needs to shift our daily practice. The Eucharist both shows us how to live in community and is a sacrament that enables us to live in community. We depend on others, others depend on us, and we all depend on God. The Eucharist shows us that relationships are based on more than commutative justice; relationships include sharing and giving.

The Trinity also echoes the idea of communion. In the Trinity, there is no hierarchy or individual prioritized. The Trinity includes God, Jesus, and the Holy Spirit, each part of the other, independent and interdependent. "The early Cappadocian doctrine of the Trinity understood the persons of the Trinity as each having their own work and character yet mutually constituted by their relationship with the others" (Snarr 2011, p. 61). God, Jesus, and the Holy Spirit are unique and work in different ways yet they are one together. "In fact, living into our nature as the Image of God (or *Imago Dei*) might be better understood as living into the Image of the Trinity or *Imago Trinitatis*" (Snarr 2011, p. 61). We often discuss humans as created in the image of God. This God is communal – three and one. So too are humans individual and in communion with others. We are each unique and part of community.

The Trinity can help us to understand what it means to be an individual in community. The Trinity exhibits how to be different and together. Of course, this concept is not easy to absorb. Articulating the nature of the Trinity and especially of Jesus' humanity and divinity has been controversial throughout Christian history, with strands of the Christian churches breaking apart over differing interpretations. Finite humans attempting to understand the infinite leads to a variety of interpretations. One such difficulty for us in the West is understanding personhood as both individual and in relation to others. "When the patristic theologians used the word *person* (*hypostasis* in Greek) ... they meant a distinct identity, an otherness, which only made sense in relationship" (Ramachandra 2008, p. 252). Part of our difficulty in conceptualizing the Trinity in the West is that we think first of an individual who might then choose to relate to another. However, we are born into relations and only learn to be an individual as we grow. Our individuality is formed in relation to community. We find our distinctive gifts as we interact with each other in community.

Freedom for the common good

Since we are inherently in relation with others, we need to consider the common good. Most simply, the common good is the good of the entire community, not one particular individual or one subset. Freedom then is not solely freedom from, but freedom to love others. We should be free to prioritize something other than the market as the center of life. We should be free to live life abundant, loving others. Humanity interrelated in unjust ways. God through Jesus shows a different path. God came to earth as Jesus to enable humans to work toward just relationships in community.

Luther articulated sin as self-interest.

> Luther understands sin as *se incurvatus in se* (self turned in upon self), the human proclivity to do everything for the promotion of self, out of concern for self, and using resources claimed as one's own rather than as gifts of God.
>
> (Moe-Lobeda 2002, p. 79)[3]

Sin is focusing on oneself, rather than the good of others. Freedom from sin means we can turn to others, care for others. With Jesus' life, death, and resurrection, a model for interacting justly was offered.

Paul argued that when Christians become part of the "body of Christ," it is as if they become part of one physical body where each part helps to make up the whole. Each part cares for the whole. We each give and receive and together we can live life abundantly.

Desiring abundant life with God and others

Capitalism wrongly argues that desire can be satiated through constant consumption. Instead, "we were created to desire God and the things of God" (Bell 2012, p. 86). Our desire is to know God and each other; however, we do not. Liberation theologians describe this misunderstanding of desire in terms of the rupture of relationships between human beings. Rather than working toward the good of all, we focus on ourselves and attempt to satiate our desires in the market. Where capitalism sees greed as a good, Christianity understands greed as sinful. Rather than aiming to accumulate, we need to understand that there is a limit to accumulation but not to relations in community.

Christian theology promotes "unconditional giving." Our lives are gifts we can share with others. "We are constituted as persons through webs of interconnectedness. We become the occasion for each other's self-fulfillment. Those who love us make us what we become; we only learn love by being loved" (Ramachandra 2008, p. 115). As we relate to others, we begin to become ourselves, to know ourselves, and to know God. We want to give of ourselves to others, in order to become fully human.

Charity, in the biblical context, is only possible in a just situation. In an unjust situation, you are simply returning to another what is already theirs. It is impossible to "give what is mine to the other, without first giving him what pertains to him in justice" (Benedict XVI 2009, p. 6). Because each person deserves to be able to live life abundantly, if I have more than I need and another does not, I cannot be charitable, I can only give to them what I have wrongly accumulated. Stan Chu Ilo, Nigerian Catholic theologian, expands, "charity demands justice both at the inter-personal levels and among communities, nations, and in religious institutions and organizations" (Ilo 2011, pp. 23–4). For us to be able to give of ourselves to others, they must already have what is due to them as human beings and we must not have more than enough. Otherwise, what we are giving is already theirs. Charity can only occur when we each have what we need as humans.

To be fully human, we need to live in a just world. A just world would enable us to be charitable. FBDOs that give time, money, and support are not doing so simply to be kind. FBDOs act, within the Christian tradition, because justice demands it. Hence, upward accountability is incompatible with the notion of justice, which is that those of us with privilege owe the poor anything we have over and above what we need.

Abundance

We can work from a notion of abundance rather than scarcity. Scarcity is a premise of capitalism because capitalism has no end point, where one has accumulated enough. What we understand ourselves as needing to be in just relation with others and God, we can then understand that our fear of scarcity is due to the sin of over accumulation. Resources are scarce because we accumulate more than we need and because we want unlimited accumulation; such behavior is sinful. God created a world where humans and nature can live together, as long as humans understand the concept of enough and that each human being deserves to live life abundantly.

It is our failure to understand enough that keeps us unhappy, anxious, and in a situation of injustice. According to Kim, we sin through our overaccumulation. "If we are to live in a harmonious relationship with others and nature we need to eliminate *han* and work toward a life which is satisfied with a fair share" (Kim 2013, p. 53). *Han* is the suffering of some caused by others. It is also the suffering we cause ourselves by doing harm. We do harm by accumulating more than we need while others go without. Christian theology understands God to have created the earth and all therein. It was "good" and it can be more than enough for each of us if we understand "enough." Living life abundantly does not include unlimited accumulation.

Human beings matter

Strands of the Christian tradition continued to argued that "all human beings are equal as equal creatures of the one God" (Gorringe 1994, p. 15). Each human being has equal value simply by being human. If people are being treated unequally, then injustice exists and we need to work toward justice. Yet, in the current global economy, although there is injustice, "the economic system was never evicted"; instead the economy evicts the poor (Althaus-Reid 2000, p. 54). As we saw, the system excludes some and exploits others and we allow this exclusion and exploitation to happen. Human beings are even considered commodities, sometimes bought and sold. Instead of people having control over the economic system, the economic system controls and excludes people.

In contrast, an economy needs to put people before growth and profit. Catholic social teaching argues that "people do not exist for the economy but the economy for the people" (Fortman and Goldewijk 1998, p. 24). In this way, Catholicism has criticized aspects of capitalism for excluding and marginalizing people. An economy needs to put the good of all people first, rather than exclude some. We will consider the economic ramifications of this theology in Chapter 5. Our current economic system focuses on its own inner workings, aiming for profit and growth. Its goal should be to ensure the survival and thriving of all human beings.

Understanding this theology, we can begin to point to God's new earth. We need to ground our economic practices in our theology, as Chapter 5 explores. This mission is not easy. The authorities killed Jesus for introducing the good news to the marginalized. To follow Jesus, we need to work toward a just new earth, where each human will live life abundant. We need to resist the assumption that the current system is as good as it gets, and is unchangeable. Instead, we need to propose alternatives that will treat every human being as human and deserving of life. As articulated by communities of the poorest:

> There is even a gospel for the rich though it may not be the one they want to hear. ... In a more demanding form it encourages the rich who look for salvation to give everything away: "on the issue of wealth the Gospel is very clear. In the words of Jesus Christ: "Sell your possessions and give in charity. Provide for yourselves... never-failing wealth in heaven"" (UN p. 7).
>
> (Taylor 2003, p. 36)

The good news is that we can live life abundantly without accumulating wealth, although those of us with privilege may not think so. There is enough for all of us to live life abundantly and justly, when we orient ourselves toward relations with others and away from accumulation. We can reorder the economy similarly so that poverty and wealth are both reduced from their extremes, in order for each human being to live.

Humans will not fully bring about the new heaven and new earth, but we can begin. This concept can be difficult for us to understand. How can we work toward a new earth when we know only God will achieve it? "In the West, time is considered to be linear; this makes it difficult to grasp biblical eschatology, where the present and the future – 'the now and the not yet' – are seen as intertwined and interconnected" (Kim 2013, p. 83). Unlike the West's conception of linear time, other cultures have more fluid concepts, which can make it easier to understand how we can be living into the new earth now. In the West, we see the past as already done, the future as uncertain, and so we focus on the present. The Spirit shows us a different concept. "The Spirit, who goes between, moves through the borders of space and time" (Kim 2013, p. 83). God, through the Holy Spirit, moves through time just as we understand movement through space. Hence, we can both be in the new earth and not yet there. The past, the present, and the future are all with us now, intertwined. We should not ignore them.

To work toward the new earth, we need to be with the poor, as the next chapter describes. Working toward justice is reframed when we think of ourselves in community, free to love others, with enough resources to meet our needs.

Conclusion

The first step to ending injustice is to dwell in injustice, acknowledge that we are living in an unjust society, and be with the poorest. The second step is to stop accepting injustice and begin to work against it. Justice spreads far beyond economic poverty and wealth, as the final chapter articulates. Injustice is any imbalance of power between human beings. To achieve justice requires right relationships between human beings.

Justice requires change by all of us within and beyond economics. In order to achieve justice, humans have to act. The central goal or value of the struggle toward a new earth is justice. This new earth, with its goal of justice, will transform relations within community, so that each of us can have abundant life.

Understanding empowerment, and the need for justice, we will now consider why those of us with privilege need to move to the side of the poor and marginalized and how to do so.

Notes

1 Culture can be defined as "a body of informal knowledge that is historically inherited, transformed, and embodied, and contested in traditions,
incorporated and innovated in practices, and transmitted, altered through
social learning, in a community of evolving and porous boundaries" (Li
2007, p. 155).
2 Note here that I am not addressing the political realm, which may work to
moderate capitalism's effects.
3 Putting self before others can also be critiqued within Islam, particularly
within economics.

> Engineer argues that greed was the first cardinal sin committed by
> humankind. ... Adam and his partner, therefore, had everything that
> they *needed* in Paradise – a place of security in which they would be
> neither hungry nor naked, thirsty nor exposed to the sun – and it was
> precisely when they coveted what was beyond their basic needs, seek-
> ing to satisfy their greed and rebelling against their Creator in the pro-
> cess, that they were banished to the Earth. ... Engineer argues that the
> Qur'an calls for a simple, even austere, style of living that is based on
> fulfilling one's immediate needs, spending all surplus wealth in the way
> of the poor and needy (Q. 2:219).
>
> (Rahemtulla 2017, p. 75)

References

Althaus-Reid, Marcella. 2000. Bien Sonados? The Future of Mystical Connections in Liberation Theology. *Political Theology*, vol. 3, 44–63. doi:10.10
80/1462317x.2000.11876941

Bell, Daniel. 2012. *The Economy of Desire: Christianity and Capitalism in a
Postmodern World*. Grand Rapids: Baker Academic.

Benedict XVI, Pope. 2009. *Caritas in Veritate*. http://w2.vatican.va/content/
benedict-xvi/en/encyclicals/documents/hf_ben-xvi_enc_20090629_caritas-
in-veritate.html. Accessed August 12, 2019.

Bradstock, Andrew and Christopher Rowland. 2002. Christianity: Radical
and Political. In *Radical Christian Writings: A Reader*, eds. Andrew
Bradstock and Christopher Rowland, xvi–xxvi. Malden: Blackwell.

Calvin, John. 1960. *Institutes of the Christian Religion*, ed. John McNeill,
trans. Ford Lewis Battles. Philadelphia: Westminster Press.

Chrysostom, St. John. 1984. *On Wealth and Poverty*, trans. Catharine Roth.
Crestwood: St. Vladimir's Seminary Press.

Claar, Victor and Robin Klay. 2007. *Economics in Christian Perspective: Theory, Policy and Life Choices*. Chicago: IVP Academic.

Fortman, Bas de Gaay and Berma Klein Goldewijk. 1998. *God and the
Goods: Global Economy in a Civilizational Perspective*. Geneva: WCC
Publications.

González, Justo. 1990. *Faith & Wealth: A History of Early Christian Ideas on the
Origin, Significance, and Use of Money*. San Francisco: HarperSanFrancisco.

Goodchild, Philip. 2009. *Theology of Money*. London: SCM Press.

Gorringe, Timothy. 1994. *Capital and the Kingdom: Theological Ethics and Economic Order*. Maryknoll: Orbis Books.

Hayek, Friedrich. 1978. *Law, Legislation and Liberty, Volume 2: The Mirage of Social Justice*. Chicago: University of Chicago Press. doi:10.7208/chicago/9780226321257.001.0001

Horsley, Richard. 2009. *Covenant Economics: A Biblical Vision of Justice for All*. Louisville: Westminster John Knox Press.

Ilo, Stan Chu. 2011. *The Church and Development in Africa: Aid and Development from the Perspective of Catholic Social Ethics*. Eugene: Pickwick.

Jacobs, Sarah. 2018. Just Nine of the World's Richest Men Have More Combined Wealth Than the Poorest 4 Billion People, *Independent*, January 17, 2018. www.independent.co.uk/news/world/richestbillionairescombinedwealth jeffbezosbillgateswarrenbuffettmarkzuckerbergcarlosslimwealth-a8163621.html. Accessed August 12, 2019.

Kim, Grace Ji-Sun. 2013. *Colonialism, Han, and the Transformative Spirit*. New York: Palgrave Pivot.

Li, Xiaorong. 2007. A Cultural Critique of Cultural Relativism. In *The Challenges of Globalization: Rethinking Nature, Culture, and Freedom*, Invited Issue of American Journal of Economics and Sociology, eds. Steven Hicks and Daniel Shannon, 151–71. Oxford: Blackwell.

Luther, Martin. 1989. The Blessed Sacrament of the Holy and True Body and Blood of Christ, and the Brotherhoods. In *Martin Luther's Basic Theological Writings*, ed. Timothy F. Lull. Minneapolis: Fortress Press.

Maimonides (Rambam). *Mishneh Torah, Gifts to the Poor*. www.sefaria.org/Mishneh_Torah%2C_Gifts_to_the_Poor.10?lang=bi. Accessed August 10, 2019.

Míguez, Néstor, Joerg Rieger, and Jung Mo Sung. 2009. *Beyond the Spirit of Empire*. London: SCM Press.

Miranda, Jose Porfirio. 1977. *Marx and the Bible: A Critique of the Philosophy of Oppression*, trans. John Eagleson. London: SCM Press.

Moe-Lobeda, Cynthia. 2002. *Healing a Broken World: Globalization and God*. Minneapolis: Fortress Press.

Novak, Michael. 1982. *The Spirit of Democratic Capitalism*. New York: Touchstone.

Pilarczyk, Daniel. 1997. *Bringing Forth Justice: Basics for Just Christians*. New York: Paulist Press.

Rahemtulla, Shadaab. 2017. *Qur'an of the Oppressed: Liberation Theology and Gender Justice in Islam*. doi:10.1093/acprof:oso/9780198796480.001.0001

Ramachandra, Vinoth. 2008. *Subverting Global Myths: Theology and the Public Issues Shaping Our World*. Downers Grove: IVP Academic.

Snarr, C. Melissa. 2011. *All You That Labor: Religion and Ethics in the Living Wage Movement*. New York: New York University Press. doi:10.18574/nyu/9780814741122.001.0001

Taylor, Michael. 2003. *Christianity, Poverty and Wealth: The findings of 'Project 21'*. London: SPCK.

Van Til, Kent. 2010. Poverty and Morality in Christianity. In *Poverty and Morality: Religious and Secular Perspectives*, eds. William Galston and Peter Hoffenberg, 62–82. Cambridge: Cambridge University Press. doi:10.1017/cbo9780511779084.005

Zohar, Noam. 2010. Jewish Perspectives on Poverty. In *Poverty and Morality: Religious and Secular Perspectives*, eds. William Galston and Peter Hoffenberg, 204–19. Cambridge: Cambridge University Press. doi:10.1017/cbo9780511779084.011

3 Being with the marginalized

Understanding empowerment and the Christian demand for justice, means we should attend to people suffering from injustice. Rather than focus on accumulating wealth, we need to focus on building just relations, particularly with the marginalized. We cannot achieve empowerment and justice until we are in relation with the marginalized. This critical piece of the Christian mission, being with the marginalized, is often missing from our practices.

> Justice ... has to do... with a particular concern for the restoration of relationship with those who are excluded ...and pushed to the margins of the covenant: the proverbial widows, orphans, and strangers of the Old Testament; and the fishermen, prostitutes, tax collectors, and the sick of the New Testament.
>
> (Rieger 2009, p. 139)

Before we can struggle toward a new earth, we need to see who is excluded and prioritize building relationships with them. The biblical texts confirm this need.

As part of the transformation toward justice, we need to prioritize the poorest. The poor are those people who lack the power to survive and thrive. The poor lack the power to live abundant life. As we will emphasize in the following chapters, the system exploits many people. Other people are excluded altogether. With the dominance of the global marketplace, people who do not or cannot participate in the market, do not count. Our value seems to be determined by what we produce and consume. I will urge us to begin with those excluded from life abundant, aiming to craft alternatives. The system works through a dominating form of power and exclusion. Those of us with privilege are also unable to live life abundant until we build just relations with others. The goal of development should be to walk alongside the marginalized and support them in the ways they ask to live life abundant, not aim to include them in our systems.

Jesus and the marginalized

Throughout the Bible, we can see God working on the side of the marginalized and calling humans to do so too. In the Hebrew Bible, God tells Moses to lead the Israelites out of slavery in Egypt. Despite Moses' misgivings, God persists in helping him until the Israelites are finally free. In the New Testament, the life and actions of Christ show the prioritization of the poor and marginalized, treating each person as fully human. From the liberation perspective,

> Jesus is not only born poor (*Jamaica*), the poor are the whole purpose of Christ coming to earth. The Beatitudes begin with his blessing on them (*Philippines*). He is open to their needs (*Fiji, South Korea*). His central message was to promise them liberation and his mission was to empower them (*Bangladesh, Chile, Germany*).
>
> (Taylor 2003, p. 32)

As the narratives of the poor articulate, the life of Jesus provides clear and consistent examples of Jesus being with and siding with the poor. As we noted in the previous chapter, Jesus worked with the poor, struggling toward justice. As Jesus' followers, we should too.

The Gospels tell us that during his life, Jesus went to the margins geographically and socially: to the poorest, where they lived. "Jesus' program was to go precisely where no politician would ever tread, where no religious leader would ever head, that is, to the villages and remote towns" (Raheb 2013, p. 104). Jesus did not live in Jerusalem with the elites; he actually did not have a home. He also did not argue with the Roman leaders. He only went to Jerusalem at the end of his life. "At the center of Jesus' attention were the ... marginalized, ...people who had to fear for their lives, people who could not walk upright because they were under so much pressure and oppression" (Raheb 2013, p. 105). While geographically at the margins, he interacted with the excluded. Jesus walked with the marginalized, announced the good news to the marginalized. To follow Jesus, we also need to walk with the marginalized, rather than remain with the elite. To bring about the new heaven and new earth, we first need each person to be empowered. Then together we can struggle to change.

As noted in the previous chapters, living and spreading the good news is a dangerous mission; it got Jesus and some of his followers killed. "Jesus' life represented a threat to all who would limit and impose an order, including a spatial and geographical order, on God's Reign. ...He 'walked with' the wrong people and in the wrong places" (Goizueta 1995, p. 203). In being with the marginalized, Jesus upset those at the center of power. The powerful usually do not want the voices of the marginalized amplified. The powerful have to repress resistance, as it causes people to realize and follow alternatives. "In Jesus' world, ... justice was defined as ensuring that every person stay in the place appropriate to him or her. To accompany the poor and the outcasts was to transgress the established and accepted boundaries which separate 'us' from 'them.'" (Goizueta 1995, p. 203) Rather than work within the system that separated people by class, religion, and so forth, Jesus moved to the margins to resist. To follow Jesus, we need to move to the margins and be with people who are excluded. In so doing, we can better see the domination of systems and structures. As we noted in the first chapter, it is difficult to understand the harm of "power over" when we are enmeshed in and benefitting from the dominant system.

To build just relationships, we need to begin at the margins, as Jesus did. We inherently are in relation with others; these relations can either be unjust or just. "As the first and foundational sacrament (*ursakrament*), Jesus reveals to us not only who God is (theology) but also who we are (anthropology) – inherently sacramental, relational creatures" (Goizueta 1995, p. 67). Our choice is not whether to be in relation with others but how to be in relation with others. We noted in previous chapters the example of the Trinity as exhibiting uniqueness in community. We become ourselves as we are in relation with others. Further, Jesus did not simply talk about the poor; he was present in their lives. This relation, to be just, requires physical presence, it is not enough to state one is on the side of the poor, one also has to be with the poor. I cannot be on the side of the poor if I do not know and spend time with poor people.

Let me make clear that being poor and marginalized is not the ideal; poverty needs to end, so that each person can have abundant life. However, to resist marginalization, we actually need to be at and work at the margins. Resistance begins with the poorest, amplifying their voices and demands.

As we noted in the previous chapter, sometimes Christ's death on the cross is seen as an atonement for human sins, as if paying back a debt. Jesus had to die so that humans could be saved. The marginalized are seen in this way in the capitalist system, wrongly. The poor atone by dying so that other people can live. Human sacrifice is unnecessary and unjust; this acceptance of sacrifice is a wrong understanding of justice and a wrong understanding of God's actions. God wants us to each live a fully human life. God sent Jesus because of God's desire for all humans to live life abundantly.

Jesus and the good news

The good news that Jesus shared is that it is possible to live life abundant. We work toward this new earth now, knowing this justice will be fully achieved in God's kingdom. God, coming to earth as Jesus, brought this good news.

> The biblical narrative reveals a God who comes to us not as master but as a servant, who stoops to wash the feet of his disciples and to suffer brutalization and dehumanization at the hands of his creatures. On the cross, God in Christ bears the indignity of all whose human dignity has been violated.
>
> (Ramachandra 2008, p. 193)

God took on the suffering of humans through Jesus and shared the news that suffering could end. We become closer to God by becoming closer to others, particularly people who suffer. As God came to be with people at the margins, so too should we.

Jesus' mission, passed on to his followers, was to spread the good news. This good news is the possibility of living life abundant and of resisting oppressive authorities. To do so, those of us with privilege need to move from our privileged spaces to the side of the poor. This shift enables us to work toward the new heaven and new earth.

> The preferential option for the poor ... is also an essential component of the prophetic proclamation of the gospel, ... that includes the connection between justice and God's gratuitous love. Working so that the excluded might become agents of their own destiny is an important part.
>
> (Gutiérrez 2013b, p. 154)

The marginalized need to be heard and to act. God's kingdom will be a place of justice, where each person is empowered and each voice is heard. As we noted in the previous chapter, we work toward the new earth even as we know only God can full bring it to fruition. God announced the new earth through Jesus, encouraging us to work toward it. The church can move back to this broader understanding.

With Jesus life and death, working toward a new earth becomes possible. Now we have a model of inclusion and love, which begins with the excluded. "The new *laos* of God incarnates not a nation but a multitude, a 'popular' people, a new experience of humanity without exclusions. Thus, ... the voice of the *laos* shows the limits of power" (Míguez, Rieger, and Sung 2009, pp. 201–2). Korean minjung theology articulates this concept of the multitude well, understanding that the masses are to rescue God from suffering. The word *minjung* (min-people, jung-mass) is similar to the word *ochlos* (multitude) in the Gospels. This multitude includes the sick, the prostitutes, the tax collectors and so forth. God is with those people at the margins, suffering with them. Toward the end of a short play by Kim Chi Ha, called *The Gold-Crowned Jesus*, a leper sees a statue of Jesus crying. He sees the statue has a gold crown of thorns and wishes he had that crown to be cured, to buy food, and to help his prostitute friends keep their home. (The play opens with a priest refusing to support the women as they try to keep their home from being destroyed.) Jesus speaks to the leper, asking him to take off the crown. He tells the leper that he has been locked up in this cement by the priests and elite. The leper fears if he takes off the crown, he will be accused of stealing. And indeed, when he does remove the crown, the priest, the company president, and the policeman all accuse the leper, while the prostitute kneels at the feet of the newly freed Jesus to worship. The play ends with Jesus again becoming cement as the gold crown is placed back on his head (1978). Currently, our world is organized around the few with money, rather than the vast majority of human beings. Privileged Christians have imprisoned Jesus. As Byung Mu Ahn, North Korean minjung theologian, argued, "The very thing that makes Jesus turn into cement is the Christology made by the Church" (1993, p. 165). It is the job of the poor and marginalized, such as the leper and the prostitute, to rescue Jesus from this prison. For those of us with privilege, it is our job to be alongside the leper and the prostitute.

This good news expands the notion of sin and salvation beyond the narrow view of it as solely individual and spiritual. Sin is injustice both in human relations with each other and in structures and systems that exclude some human beings. God calls human beings to repent from sin. De la Torre develops the notion of conversion to the poor, as Gutiérrez did, in terms of salvation and sin, with reference to Juan Luis Segundo, Uruguayan priest and liberation theologian.

If conversion is understood as a rupture with and a turning away from sin (sin caused by individual actions *and* sin caused by social institutions), then salvation can only occur through the raising of consciousness to a level that can recognize the personal and communal sins preventing the start of a new life in Jesus.

(de la Torre 2004, pp. 42–3)

Repentance demands a change in heart and in action, a start in a new direction, following Jesus. It means recognizing how broad and deep sin is. The first chapter argued how structures tend to use power to dominate rather than empower. The second chapter explored self-interest, desire for unlimited accumulation, and an economic system that sacrifices some humans so that others can live. We need to turn away from these sins and work toward a new heaven and a new earth.

Conversion and salvation are processes not momentary events. One is not "saved" and then done with sin. Each step toward the new heaven and new earth takes work. "A conversion is the starting point of every spiritual journey. ... "Sell all that you have... and come, follow me" (Lk 18:22). ... Without this second aspect the break would lack the focus that a fixed horizon provides and would ultimately be deprived of meaning" (Gutiérrez 2013a, p. 71). With belief comes action, praxis: a combination of action, reflection, and action again. It requires giving of ourselves to the poor. In order to experience salvation, one needs to walk with Jesus at the margins, learning from others at the margins, and learning how simply to be with others at the margins.

After conversion, sin continues to exist in our lives. Conversion means we are aware of sin and working to resist it. Sin is with us, just as the new heaven and new earth are with us. Both are consistent options. "To sin is to deny love.... Many of those who are committed to the poor freely admit the difficulties they have, as human beings and believers, in loving God and neighbor" (Gutiérrez 2013a, p. 73). Here I want to introduce the concept of backsliding (a popular term in Pentecostal churches). Conversion to the poor is not a final or static concept, as some people assume. Backsliding is the process when people slide back toward where they started, back to the "immoral" life they led before become a Christian. There is no guarantee after conversion to the poor that a person or a community will move forward consistently. We need to continue the work.

In Christian theology, the Holy Spirit guides us, helping us to understand and navigate our options, as we try to work toward the new earth. The Holy Spirit walks with us and is within us as the breath of life. "It is God's Spirit which lives in us and dwells in our being and our lives which will bring about change" (Kim 2013, p. 70). We can choose either to listen to or ignore the Spirit. But if we want to change, we need to listen. The Spirit empowers us as it empowered the early church. "The Spirit is so powerful that it can alter our lives and literally invert the way we see things" (Kim 2013, p. 71). If we listen to the Spirit, we can see a path toward a new heaven and a new earth, and we know that God is present with us; we can propose alternatives to premature death.

Forgetting the good news

In the previous two chapters, we saw how, with the expansion of the early church among the wealthy and particularly with the conversion of the emperor Constantine in 330, (1) the understanding of power as empowerment shifted to a dominant form of power and (2) the understanding of justice, poverty, and wealth as meaning anything you accumulate above what you need actually belongs to another in need shifted. The notions of charitable giving and accumulation as a good emerged. Here, the understanding of sin and salvation began to narrow, as did the concept of mission. Gorringe argues, "'mission' originated in the resurrection and Pentecost: it was about spreading the good news of the death of death, of the overcoming of alienation, of the possibility of a new type of human community" (Gorringe 2004, p. 256). As we saw above, Jesus' mission was to empower the marginalized and to resist the powerful. For enacting this mission, Jesus was crucified by the powerful. However, the narrative continued because Jesus was resurrected. This message was a powerful motivator for Jesus' followers. It was possible to overcome the dominant powers and work toward a new heaven and new earth.

Gorringe articulates three ways the Christian church then failed to carry out this mission. First, rather than keeping a non-hierarchical community as an alternative vision of the new earth, the church consolidated itself into hierarchical bodies and sided with the powers that be: emperors, monarchs, and states through Christendom and colonialism (Gorringe 2004, pp. 256–7). As noted in the first chapter, it upended the notion of a good use of power as empowerment and turned Jesus and God into a divinity that used and supported "power over."

Second, not only did the church behave in a way opposite to its mission, it also re-interpreted the mission. Gorringe argues the church reduced the understanding of salvation to individual repentance from individual sins. Sin narrowed from "the complex ways in which human beings destroy themselves and God's good creation..." (Gorringe 2004, p. 257). In order to be able to see themselves as Christian while enacting a dominant form of power and accumulating wealth, the church had to redefine the good news. The good news became about an individual's relationship with God, not with others. In some parts of the Christian tradition, salvation was limited to "accepting Jesus as one's savior." One was to live a holy life as an individual but any relation to justice in community or to structures disappeared.

Finally, the Church separated the notion of "proclaiming the good news" from the daily life of the church. The church became an institution with several parts, only one of which was "mission-oriented." That mission was narrowed to trying to convert others to Christianity, through individual salvation. The churches added "boards of mission." Gorringe argues, this shift "led to the loss of the perception that the Church itself was missionary, that the living... of the good news of the divine flesh taking was simply the nature of the Church" (Gorringe 2004, p. 257). The members who made up the church were no longer models of the good news. Instead, the church sent particular church members out to spread the narrowed version of the good news, which could somehow be separated out from the lives of individual Christians and the structure of the church itself. It was enough for a Christian to simply tell others about the good news, they did not have to live it out.

Hence, today, there are still Christian development organizations that argue that "mission" in terms of individual salvation has to be part of development work, rather than understanding that mission is development work if it is accompanying the marginalized and working toward justice and empowerment. Mission and salvation are misunderstood. Jesus' message included ending injustice.

Mission: accompanying the marginalized

God is on the side of the marginalized, not because the marginalized are better than anyone else but because God empowers the powerless. Roberto Goizueta, Latino Catholic theologian, states, "If 1) God's love is universal and gratuitous..., and 2) God's love is made manifest in history, and 3) that history includes injustice..., then 4) God's love *must* take sides with the victims of that injustice" (Goizueta 1995, pp. 175–6). Just as marginalization is not an ideal, so too the poor are not the ideal. God sides with the poor because the poor suffer injustice. We are also to side with the poor, not because the poor are so much better than those of us with privilege but because we need to build just relations. God loves us all equally but does not side with us all equally. God sides with people experiencing harm.

It is common to hear Christians ask "What would Jesus do?" A better question to orient us is "what would we do to Jesus?" Jesus is the "other" with whom we are in relation. As Gregory of Nazianzus stated: "Let us take care of Christ while there is still time, let us minister to Christ's needs, let us give Christ nourishment... let us give this gift to him through the needy" (Gregory of Nazianzus 2003, 14:40). When we are with the marginalized, we are with Jesus. To ignore the suffering of someone is to ignore the suffering of Jesus. God is found in our treatment of the poor. "For I was hungry and you gave me food; I was thirsty and you gave me something to drink... Truly I tell you, just as you did it to one of the least of these who are members of my family, you did it to me (Matthew 25:35–6, 40). Jesus identified himself with the poor. As Christians do to the poor, they do to Jesus. Hence, when there are unjust relations between humans, we are treating Jesus unjustly. The notion within some development practice that poor people are "backward" or "less than" is turned on its head when we think of the other as Jesus.

Those most marginalized are those who have died prematurely, sacrificed in our current system. In the New Testament, Jesus raises people from the dead, returning them to life, Lazarus, for example. In our context, we cannot raise the dead. However, it is important to begin our questioning of the situation with those who have died and those people who are barely surviving. Who has died and why have they died?

> Archbishop Romero... said: Lord, this day our conversion and faith draw support from those who lie there in their coffins. They are messengers who convey the reality of our people and the noble aspirations of a church that seeks naught but the salvation of the people.
>
> (Gutiérrez 2013a, p. 82)

Within many cultures, the dead also remain part of the community as ancestors; they are present. Beginning with those people who have died too soon can help us to see the margins. The poor often die to enable the privileged to keep living as they do. We need to end this sacrifice.

So how can we learn to treat all human beings as human? The first step, which seems simple although it often eludes us, is to recognize that each person is a human being. In our current economic and political systems, we only enable people to access the basic resources they need for life if they have money and we prevent people from moving to safer spaces where they can access resources.

Second, as human beings, we need to recognize that we are in relation with others. In the African context, one of the concepts used is *Ubuntu*. "People exist in a web of relationships...: *umuntu ngumuntu ngabantu* (Zulu: a person is a person through other persons). ...If one person's dignity is violated, it is said that other people are also affected" (Hankela 2014, p. 2). One cannot recognize another human being simply through words; it takes action in community. If we fail to recognize others, we fail to be fully human ourselves. We cannot be in communion with God without being in communion with each other.

One is a human being and is also becoming or being kept from becoming human, depending on community. "A human being becomes rather than is" (Hankela 2014, p. 52). Our humanity is enacted through our relations with others. "A human being... is perceived as existing and becoming in a community" (Hankela 2014, p. 54). One becomes human by treating others as human. One becomes human by being treated by others as human. This connection happens in community. In Johannesburg, South Africa,

> the Refugee Ministry aimed at making the dwellers understand that... their intrinsic value as humans – was something that other people could not deprive them of. ... [Bishop Paul] Verryn also addressed human dignity as a characteristic of a human being (or human community) that needs to be actualized.
>
> (Hankela 2014, p. 171)

In the situation of southern African migrants to South Africa, while migrants are already human beings, they need to be treated as human beings in order to be able to live life abundantly. Currently, migrants' humanity is often denied, as the section on xenophobia in Chapter 6 explores; that chapter also details several other ways people are excluded and considered less than fully human.

Before we even make a choice, we are in relation to others. The easiest way to see this interdependence is to think about being born. One does not magically appear in this world as a fully formed individual. We are in relation with another human being before being born. Second, after birth, we are dependent on others for a significant amount of time. Hence,

> to love the poor preferentially is to make an *intentional* choice to be that which we already *are*...: individual persons defined by our

a priori relationships to others, to humanity, to the universe, and, ultimately to the Triune (i.e., intrinsically relational) God.

(Goizueta 1995, p. 179)

In our daily lives, we already have relationships with people at the margins, for example, through the goods we purchase, environmental resources we use, and so forth. However, these relations are usually unjust. Our choice is whether to continue injustice or whether to struggle toward justice by accompanying the marginalized.

We work toward the new heaven and new earth so we can each live life abundant. To be fully human, we need to ensure our relations with others are just. "Relationship is not something that 'happens to someone...'; it is something one *does*...; through relationship, we discover and live our out identity as intrinsically relational beings" (Goizueta 1995, p. 72). We become ourselves as we interact with others; how we treat others shows us who we are. When we treat others unjustly, neither they nor we can live life abundant. Relations with others already exist: these relations can either be just or unjust.

The first step in rebuilding just relationships is to see who we are ignoring or excluding. Although we are already in relation with people at the margins, these relations are usually hidden. We do not see people at the margins as our neighbors. We often also "choose" who we see as our neighbors, excluding some who are geographically close to us.

In our neighbourhood are children, persons with visible mental and physical disabilities, persons who are homeless and landless and nameless. ...They are persons we do not need and who, in our view, are dispensable – or even worse, a burden to the neighbourhood.

(Oduyoye 2004, pp. 54–5)

The margin is not simply a physical space outside our daily existence. People who we pass by everyday can be invisible to us. We need to see them, be with them, and love them. First, we need to see who is actually around us. For example, who do I walk past on the street without noticing? Then, we need to see who we are in relation with who we have not paid attention to. One such example would be the person who has harvested the apple I am eating and the person who sewed the clothes I am wearing. Both of these areas are blind spots for us.

God is with the poor and so we must be too. An understanding of God comes from the poor. It is a very different starting point from a place of privilege. Much theology, as I have shown in the previous two chapters, post-Constantine, came from a place of relative power and privilege. Theology that begins with the poor will look different to theology that begins with the rich.

> When the theologian places him or herself in this particular place (the ghetto, the barrio, the public square) he or she soon discovers that the randomness and chaos of "the city" are not abstractions, but are the common lot of poor *persons*: random illness, random death, random gunfire, random searches, random employment and unemployment, random health care, etc., etc. To abandon the poor to these instruments of death while the privileged few escape into geographical isolation is to condone and perpetrate genocide – for [they] …are, of course, predominantly poor African American and Latino children.
>
> (Goizueta 1995, pp. 200–1)

In particular, being with the poor, one immediately sees that evil is not an occasional occurrence but regularly experienced, as explored in the first chapter. The spiral of violence begins with hunger, thirst, and lack of safe shelter. The question is not about why bad things happen. It is obvious to people at the margins that those of us with privilege harm them. The question is how to stop this harm. In order to understand this perspective, we actually need to walk alongside the poorest. This choice is not easy. To walk with another amongst violence, hatred, sickness, etc., is a difficult choice but that is where Jesus went and where Jesus is now.

In contrast, those of us with privilege prefer to avoid the poor. Of course, the poor may prefer to avoid the poor as well. Growing up poor, all I wanted was escape my neighborhood for something better. I wanted to study poverty but not remain poor.

> As a society, we are happy to help and serve the poor, as long as we don't have to walk *with* them where they walk, that is, as long as we can minister *to* them from our safe enclosures. The poor can then remain passive objects of our actions ….
>
> (Goizueta 1995, p. 199)

This distance from the poor, unfortunately, enables structures of sin to continue and prevents us from living abundant life. Donating money to one's favorite charity is not the same as accompanying the poor. Whilst giving away anything more than what you need is a requirement of justice, it is also important to actually build relationships with people at the margins to struggle with them.

We cannot have a just relationship with God, if we fail to have just relations with each other. In our globalized world, this requires care and attention to each human being, as our actions impact others across the planet. It is in learning to love each other that we learn to love God; it is in knowing each other that we know God. It is not possible to love God without loving each other. As I noted in the first chapter, Jung Mo Sung, a Korean-Brazilian economic theologian, states that when we are able to see an old, ill, poor, lesbian sex-worker as made in the image of God, and equal in value to ourselves, then we are beginning to opt for the poor (Sung 2005, p. 5).

We need to love each human being, simply because they are human, regardless of appearance, characteristics, or what they can do for us. "The clearest criterion for judging whether...human praxis has been instrumentalized is that society's treatment of its most 'useless' and 'impractical' persons, for example, the elderly, children, the handicapped, the poor, the unemployed, the terminally ill, the homeless" (Goizueta 1995, p. 108). Our society tends to value people in terms of what they can contribute, rather than considering each human being to be of equal value. We exclude and discard many lives that we consider useless. Instead, God values humans because they are human. So too should we. Each human being is created in God's image. Each person is Jesus.

The key to conversion is to simply be with "the least of these," those people who have been excluded.

> Accompaniment of the lonely poor involves walking with – not behind or in front – but beside a real person on his or her own particular journey in his or her own particular place and time, at his or her own particular pace.
>
> (Block and Griffin 2013, p. 6)

We are not to lead or guide but simply struggle with. We often jump in wanting to help. But the first step is to simply be with, and begin to know others. Unlike a notion of development as a project that can solve a problem, Christian development practice is simply to walk alongside the poor and to support their aims. Being with is already working toward justice. Struggling with is another step.

> The notion of "walking with"... is, first of all, a concrete, physical, historical act. ...One walks in a particular direction. That directionality implies, in turn, an ethical-political content: ... "Who determines the direction?" ...The possibility of accompanying the

poor does not exist unless and until the poor themselves are equal participants in dialogue and interaction.

(Goizueta 1995, pp. 207–8)

Walking with the poor means letting the poor determine where to go and struggling alongside them to get there. We need to let those who are marginalized decide how they can live life abundantly. Walking with the poor is a pre-condition for justice. Walking with means meeting, getting to know, and journeying with.

It is critical that in walking with the poor, we acknowledge that each person is equally made in the image of God. One human is not worth more than another. Accompaniment "'suggests going with another on an equal basis,' and, thus, implies the transgression of discriminatory barriers" (Goizueta 1995, p. 206). Without this acknowledgement of equality, we perpetuate structures of sin and unjust relations. When we accompany, we value the other as much as ourselves. It is in knowing the other that we come to know ourselves and God. We are equal partners in the journey toward justice.

With this equal walking with, we are better able to become fully human, freeing ourselves from the need either to take charge or to submit. We can simply be with.

That relinquishment of control in order to enter into a genuine union with another becomes, paradoxically, the place where we discover our subjectivity, our freedom as unique persons capable of *inter*acting with other persons. This freedom is at the same time a freedom *for* relationship and a freedom *from* the need to control the other.

(Goizueta 1995, p. 109)

Letting go of power over frees us and the other. We are now free to love the other simply because they are human, because we are each created in the image of God.

Being with the marginalized takes work and commitment. It is not a simple task. Those of us with privilege carry a lot of baggage, literally and metaphorically. "Such a liberative social ethic of the complicit must work toward an understanding of what it means to be complicit-allies. ... This will require serious analyses of identity and power" (Kirk 2013, pp. 159–60). When we walk alongside the poor, we will learn their perspectives. If we amplify the voices of the poorest and allow them to shift the narrative, then change is possible. The next chapter begins to explore what this accompaniment could look like in practice.

I want to end with one caveat for those of us with privilege. One can only accompany the marginalized if they say yes. Solidarity, of course, is incredibly difficult for those marginalized to accept from people who are privileged. "Solidarity, from positions of relative economic privilege, is solidarity with people brutalized by the very ideologies and structures providing that privilege" (Moe-Lobeda 2002, p. 119). Trust takes time to build and cannot be forced. One of the aspects discussed in the following chapter is how to actually be in solidarity, which requires addressing power, as the first chapter discussed. We need to remember that conversion to the poor is an ongoing process. We are always learning how to accompany people at the margins.

Conclusion

Converting to the poor and accompanying the marginalized requires a major shift in the lives of development practitioners and the organization of many FBDOs. The following chapters offer a few starting points for practice. First, we need to find ways to be with the marginalized, whether in our own communities or abroad. Second, we need to amplify the voices of the marginalized to the powerful, advocating change. True conversion to the poor would change the daily lives of Christians and the life of the Christian church.

References

Ahn, Byung-Mu. 1993. Jesus and People (Minjung). In *Asian Faces of Jesus,* ed. RS Sugirtharajah, 163–72. London: SCM.

Block, Jennie Weiss and Michael Griffin. 2013. Introduction. In *In the Company of the Poor: Conversations with Dr. Paul Farmer and Fr. Gustavo Gutiérrez,* eds. Michael Griffin and Jennie Weiss Block, 1–14. Maryknoll: Orbis Books.

De La Torre, Miguel. 2004. *Doing Christian Ethics from the Margins.* Maryknoll: Orbis Books.

Goizueta, Roberto. 1995. *Caminemos con Jesús: Toward a Hispanic/Latino Theology of Accompaniment.* Maryknoll: Orbis Books.

Gorringe, Timothy. 2004. *Furthering Humanity: A Theology of Culture.* Burlington: Ashgate.

Gregory of Nazianzus. 2003. *Select Orations,* trans. Martha Vinson. Washington, DC: Catholic University of America.

Gutiérrez, Gustavo. 2013a. Conversion: A Requirement for Solidarity. In *In the Company of the Poor: Conversations with Dr. Paul Farmer and Fr. Gustavo Gutiérrez,* eds. Michael Griffin and Jennie Weiss Block, 71–93. Maryknoll: Orbis Books.

————. 2013b. The Option for the Poor Arises from Faith in Christ. In *In the Company of the Poor: Conversations with Dr. Paul Farmer and Fr. Gustavo Gutiérrez*, eds. Michael Griffin and Jennie Weiss Block, 147–59. Maryknoll: Orbis Books.

Ha, Kim Chi. 1978. *The Gold-Crowned Jesus and Other Writings*. Maryknoll: Orbis Books.

Hankela, Elina. 2014. *Ubuntu, Migration and Ministry: Being Human in a Johannesburg Church*. Boston: Brill. doi:10.1163/9789004274136

Kim, Grace Ji-Sun. 2013. *Colonialism, Han, and the Transformative Spirit*. New York: Palgrave Pivot.

Kirk, Jeremy. 2013. Popular Messianism, Complicity, and the Continued Relevance of Liberation Theology. In *The Reemergence of Liberation Theologies: Models for the Twenty-first Century*, ed. Thia Cooper, 153–63. New York: Palgrave.

Míguez, Néstor, Joerg Rieger, and Jung Mo Sung. 2009. *Beyond the Spirit of Empire*. London: SCM Press.

Moe-Lobeda, Cynthia. 2002. *Healing a Broken World: Globalization and God*. Minneapolis: Fortress Press.

Oduyoye, Mercy. 2004. *Beads and Strands: Reflections of an African Woman on Christianity in Africa*. Maryknoll: Orbis.

Raheb, Mitri. 2013. *Faith in the Face of Empire: The Bible through Palestinian Eyes*. Maryknoll: Orbis.

Ramachandra, Vinoth. 2008. *Subverting Global Myths: Theology and the Public Issues Shaping Our World*. Downers Grove: IVP Academic.

Rieger, Joerg. 2009. *No Rising Tide: Theology, Economics, and the Future*. Minneapolis: Fortress Press.

Sung, Jung Mo. 2005. "The Human Being as Subject." In Ivan Petrella, ed. *Latin American Liberation Theology: The Next Generation*, 1–19. Maryknoll: Orbis Books.

Taylor, Michael. 2003. *Christianity, Poverty and Wealth: The findings of 'Project 21'*. London: SPCK.

Part 2

Practice

The second half of this book considers what must change for the powerful to be with the powerless as they struggle for justice.

4 Empowerment in practice

FBDOs need to move away from "projects" and toward accompanying the poorest, supporting their local efforts, and advocating for international change. This change in action requires shifting power relations between donors, practitioners, and the poorest. Paul Farmer suggests we consider aid as accompaniment. For Farmer, accompaniment "would call us to identify and support institutions that truly represent the poor; partner with public institutions rather than "go it alone"; ... invest directly in the poor themselves; and establish learning loops that put into practice what we learn through this accompaniment process" (Reifenberg 2013, pp. 194–5). The previous chapters argued that for development practice to be Christian-infused, it needs to empower the poorest, expand its notion of justice, and accompany the marginalized. This chapter begins to address how development practices could shift to accompany the poorest.

Much "normal" development work centers around people from the global North attempting to "improve" the global South. Explicit in this type of development is that the global South is less "advanced" than the North. Working from this mindset, it is impossible for power sharing to occur. "Partnership presupposes equals, no matter how different the equal parties may be" (Freire 1975, p. 39). To be with the poorest and work with the poorest, we must first see the poorest as our equals. Thinking we have the answers for the poor leads to further oppression, ignoring much of the injustice.

Development studies proceeds in a way that makes practitioners think they have the answers. Rather than training in ethics, facilitation, participation, and intersectionality, courses in economic and political development tend to be core with options on gender, migration, security, agriculture, etc.[1] This core can lead us to treat the poor as objects rather than subjects and focus on one narrow area rather than the intersections of marginalization. The difference, according to Freire, between a person as subject or object is this: "the integrated person is person as subject," while "the adaptive person is person as object" (Freire 1974, p. 4). Do development workers treat the whole person, accepting them as they are, or do they expect them to adapt to the plans of the development worker? This cycle of objectification must end.

Toward a model of empowerment

The question of power and empowerment arises with the participation of the marginalized. The changes in the situation of the powerless require a change in attitude from both the powerful and powerless. These changes encompass all realms.

The poor put the issue of power into their own definition of well-being. I define empowerment in three parts: (1) the recognition of a lack of power, (2) becoming aware of the wider society, and (3) becoming active in that society, being able to control one's own life. "Conscientization is the deepening of the coming of consciousness... [requiring] the indispensable seriousness of wanting to know rigorously" (Freire 1993, p. 109). With conscientization, people realize they are subjects rather than objects. They can act rather than be passive. Human beings can recognize their place in the world and how religion, culture, politics, and economics interact to cause this situation. The process of conscientization continues as subjects learn how to interact with the surrounding world.

Freire shifted his emphasis from one aspect of empowerment to another over the course of his life. In his earlier work, he emphasized the poor realizing their own situation of disempowerment, becoming aware of the outside world, and then acting within that world. However, over time, his emphasis shifted. While still retaining hope that structural changes would occur, he shifted his focus onto the educator/student relationship, as an introduction to the broader process. Freire emphasized the first definition of empowerment in his earlier work, particularly *Pedagogy of the Oppressed* (1970). Later, he explored how the process of conscientization could be introduced, while retaining a hope that the poor would still come to self-realization. In the educator/student relationship, students can learn for themselves with the educator's guidance. Freire believed a person could not learn alone about the processes of society. However, people can be given the tools to learn by the educator, in a community setting. Thus, Freire emphasizes a balance of power between the educators and the students. One possible role of the development practitioner is to connect local knowledge to global knowledge.

Freire argues that the powerful must become active participants in this process. Freire calls this "conversion to the people," as we explored in the previous chapter (Freire 1970, p. 47). This conversion requires the powerful to end the use of "power over." This step is often missed by development workers. Freire acknowledges that this part is difficult to achieve. Most development workers focus on alleviating poverty, ignoring the excessive wealth and dominance of the North.

Chambers focuses on the third aspect of empowerment, taking control of one's surroundings, through the characteristics of Participatory Rural Appraisal (PRA). The goal in this form of empowerment is for the outsider to pave the way for the insider to control their own lives. The outsider actively disempowers themselves and others with privilege, so the poor can empower themselves through everyday actions.

Empowerment means, for Chambers that insiders gain control in two realms. First, insiders must gain "ownership and control of productive assets" (Chambers 1993, p. 11), economic power. Second, insiders must be able to participate in their governments and "make demands on government bureaucracies" (Chambers 1988, p. 52), political power. In the final chapter, we consider other ways power and powerlessness intersect.

The marginalized must become aware of the wider world, particularly in terms of the pressures they face. Thus, Freire's method of conscientization is crucial for a mixed model of empowerment and should form, with the aspect of self-realization, a basis for empowerment. A model of empowerment should also include the poor acting to change their own lives, not simply learning how to act. Further work is needed on the macro-level in terms of ensuring the participation of the poor, not just in small local processes but amplifying these processes within the wider world.

What the poorest articulate is that they want people to join them in the struggle, beside them, not above them. An Institute of Development Studies, UK, study showed "The poorest and most marginalized were more concerned with *how* development was delivered, than *what* development was delivered" (Burns and Worsley 2015, p. 3). How the privileged interact with the poor is critical. Treating each human as fully human is the starting point for interacting with the marginalized.

In the system of methods and behavior called PRA insiders should control every stage within a development project from naming the problem to critiquing the results. Thus, any methods used aim to help outsiders let go of power over and empower the insider. The act of praxis is more important than the result of that praxis. In PRA, practitioners "have roles as convenors, catalysts and facilitators" (Chambers 1997, p. 108). In PRA praxis, reflection on power relations and behavior lead to empowering action.

Basically, a balance should be achieved whereby the reliance on the community is more important than reliance on the facilitator. First, self-realization (the first part of empowerment) would help emphasize the community aspect. Second, developing this community reliance would help to create more possibilities for facilitation. For example, Freire advocated a team of rotating facilitators. Perhaps each person within a community could take on this role. Finally, the development of a macro-level response by the poor themselves would aid in the shift away from reliance on the facilitator alone.

Participatory processes need to include the poorest at all levels. The community together will articulate the questions to be answered or problems to be solved, figure out a way to proceed and then act and reflect. The community itself decides each state. In this way, learning and progress occur.

What holds us back from achieving the empowerment of the poorest?

Institutions

Since the recession in 2008, aid has been reduced and donors are increasingly concerned to know that their money is used efficiently. There are several problems with this control, not least of which is that the money belongs to the poor, to restore justice. It does not belong to the rich. However, donors understand the money as theirs or their constituents' in the North. The language in vogue is "value for money" (VFM).

> Our bargain with taxpayers is this. In return for contributing your money to help the world's poorest people, it is our duty to spend every penny of aid effectively. My top priority will be to secure maximum value for money (Andrew Mitchell, 2010).
>
> (Shutt 2015, p. 58)

Note here the assumption that the money belongs to the taxpayers and it is optional to help the poor. Further, when money is given, strict control is kept to ensure it is used in the ways those taxpayers would approve of. FBDOs should reject this framework, articulating that justice calls for giving without requirements, since the money truly belongs to the poorest. (This need for redistribution is also why loans are inappropriate.) Instead, FBDOs work within this frame. If an FBDO wants to address value for money, it should come solely from the perspective of the poorest, who determine how the money should be spent and what makes it effective.

Even within the current system of the donors assuming they need to control funds, there are serious problems. First, the requirements for reporting come from the top down and often do not match the reality of the field. Hence, when reporting, staff are frustrated in their attempts to describe what is actually happening or staff simply report what they assume the donor wants to hear. Second, the reporting requirements are onerous, which means that staff spend time reporting rather than working in the field with the communities. Third, donors decide how the money should be spent and staff in the field have to implement what is decided. Ola Abu Alghaib, Palestinian disability rights activist, argues that "most donors behave as if societies were predictable machines and that change occurs through cause-and-effect processes attributable to a given intervention" (Alghaib 2015, p. 115). Her explanation of the hoops necessary to go through for funding and the repeated requests from the donor for re-editing the forms are enlightening. At the very least, rather than the donor requiring this organization to hire a consultant to help staff fill in the forms, the donor should be present, to see how the goals of both organization and donor could be met. Any donor requirement (there should be none) should be at the very least facilitated by and paid for by the donor. Alghaib suggests: "If donors are to contribute to real social change and empowerment, they ... need to keep their application processes as simple as possible, simplify their reporting requirement, and adapt communication methods to increase accessibility" (Alghaib 2015, p. 129). Rather than putting the onus on people to prove how they will use the money, donors should make gifts open-ended, available whenever requested, and for the long-term.

Further, with attention to "good governance," donors distrust governments that do not meet the set standards. When aid goes to the poorest, it often goes into unstable situations where results cannot be predicted or easily measured. Situations and needs will change over time.

Finally, there is less space in budgets for staff to be with the poor. Some organizations have cut travel and limited staff to staying in the city in which they work (Hinton and Groves 2004, p. 14). What is needed is more time with the poorest in community.

Donors and development organizations need to realize that the poorest should determine how and when development is succeeding. Donors should build long-term relations with the poorest communities, responding to community demands, listening to the community and understanding that change will be a complex process often without specific measurable goals (Burns and Worsley 2015, p. 76; Balboa 2018, p. 50). Donors need to be accountable to the poorest, letting go of power, and understanding that improving the lives of the poor requires more than money.

Where practitioners have implemented participatory development in many places, donors have kept the same requirements. Their agenda is to fund as simple and results-oriented a project as possible. Instead of agreeing to support a particular community of the poorest and responding to their requests, donors require specific information up front and want to disburse large amounts of money that will clearly bring quantifiable benefits in a short time frame.

FBDOs need to articulate clearly to donors that the money they give already belongs to the poor. Further, they need to articulate that what is important is the relationship that can be built between the poorest communities and the donor, which can be mutually beneficial. We need to be accountable to the poorest and the poorest should decide how to measure that accountability. A good start from the donor perspective would be asking four simple questions and letting the community know that funding and support will be there for the long-term whatever the answers are (1) What do you need? (2) How can we help you? (3) Is our help working? (4) What should we change?

Currently, much NGO development practice is seen as harmful rather than helpful. George Ritzer, U.S. sociologist, lists some negatives associated with international NGOs, including their tendency to be non-democratic and elitist: "Perhaps the strongest criticism of INGOs is that they "can be seen as neo-liberalism's 'Trojan horses' furthering its agenda while seeming to operate against some of its worst abuses" (Ritzer 2009, p. 165). NGOs need to ensure that they are working with local communities to amplify their voices, nationally and internationally, so structures can be changed. If NGOs only seek to provide services, then the NGO becomes "an ideological and political tool used by capital to reproduce itself... On this land that is the mother of liberty [Haiti], NGOs zombify the poor masses" (Lwijis 2012, pp. 71–2). Haiti is an excellent example of NGOs working completely outside the government and often in competition with each other. Haiti has thousands of NGOs failing to enact sustainable change. We will discuss one exception to this majority in the final chapter: Partners in Health.

This necessary shift requires rethinking each level of the process of development. For example, even assessment of development organizations themselves are rigid and based on measurable metrics. Christina Balboa, U.S. scholar of organization theory and governance, notes that websites evaluating NGOs, for example, Charity Navigator, include growth as an indication of NGO success (2018, p. 10). Growing an NGO is not critical but an NGO enabling local communities to assess and work toward changes and ensuring these perspectives are amplified at a global level for structural change are both important.

Practitioners

First, development practitioners need to recognize that they have biases, whether implicit or explicit. Chambers has developed a list of biases: spatial, where people tend to visit places easily accessible; project, where people visit places where projects already are; seasonal, where people visit during dry seasons; security, where people stay in safe areas or avoid outbreaks of diseases; urban slum bias, where people visit the famous slums; and airport, where people visit areas within close proximity to the airport. Three biases are less common now than earlier on: person, where people want to meet village elites; diplomatic, where people avoid sensitive subjects; and professional, where people tend to focus on their own area of expertise (Chambers 2008, p. 45; Chambers 2017, pp. 30–1). In each case, the poorest are excluded.

Note that this list of biases is for practitioners who do travel to sites to attempt to see the poor! Imagine what blind spots and biases exist for those of us who sit behind a desk and avoid the poorer neighborhoods even in our own communities. We need to change our behavior.

To counteract blind spots, practitioners need to spend time in the poorest communities. One way is the "reality check" approach or "immersion." The understanding of the importance of immersions is growing. Chambers cites a 2010 Conservative party Green Paper that stated:

> DFID staff in poor countries will spend a week living with a poor family, sharing their experiences, listening to their views and learning from their insights. Senior London-based members of DFID will also be expected to undertake such immersions. We are working with the world's poorest; we must understand their lives in order to serve them well.
>
> (Chambers 2012, pp. 172–2)

Immersion helps to begin to build relationships with the poorest. This Green Paper advocated immersions even for senior staff at DfID, which all organizations should implement in the poorest communities.

For immersion, a group of people travel to a particular community, are oriented to the community, and then spend a few days and nights in the community each living life with a different family. The visitors work, eat, and sleep with this family, observing, helping out, and talking. After the few days, each guest then meets up with the larger group to compare their experiences. In Bangladesh, participants identified several benefits: "depth, respect for voice, flexibility, and simplicity – and an orientation of learning rather than finding out (Pain et al., 2014)" (Chambers 2017, p. 135). While immersions are short term, they do help development workers better understand local situations. Immersions should occur regularly.

It would also be helpful to set up reverse immersions where the poorest spend time in communities in the global North. If just relationships are to be built, they cannot be limited to one space and to one side traveling to the other. Such immersions would be eye opening for both parties.

Development practitioners do not lose power when they empower others. First of all, it enables learning. Second, it is effective. Third, it relieves the stress of decision-making (Chambers 2012, p. 159). Practitioners need to let the marginalized become empowered, walk with them and help remove the barriers that keep them powerless.

One tension within development is that while the goal of development is the end of a need for development, on the other hand, people have careers as development practitioners and would like to continue these careers. Another tension is the goal of progress in a career, which means pleasing the boss rather than the poorest. For example, Jeffrey Haynes, UK professor of politics, states that while managers want large budgets, smaller sums of money, disbursed when a community needs them, are much easier for a community to handle (Haynes 2007, pp. 106–7). Money can be helpful but huge sums are not a panacea. In fact, large sums can be detrimental, if not spread amongst the poorest who are enabled to use the money for change.

How to be with the poorest

Development practitioners need to work with and be with the poor, not work for the poor. Being with the poor is to experience life with them, understanding their situation. Working with the poor is to accompany them as they make decisions about how to improve their lives and supporting them in these decisions. This work includes amplifying their voices in the global North to advocate for change.

Practitioners need to understand complexity theory and pay attention to the intersections, as the final chapter explores. "In complexity theory the social landscape is made up of ideas or positions or activities around which people and activity are centred" (Burns and Worsley 2015, pp. 27–8). In complexity theory, one understands that progress is not step by step with each step producing a clear measurable change. Change works by creating more attractive options and unmasking the hidden norms. It can feel like nothing is happening. However, Burns and Worsley note that in this system tipping points occur. A tipping point is when all the small attempts to change suddenly enact change, adding up to more than the sum of their parts (Burns and Worsley 2015, p. 28). In order to enact change, one needs to continue to support these small attempts. "We need (a) to ensure that there is diversity at a local level and that any inquiry encompasses the diversity that exists within the system and (b) our focus needs to be on the most local level of interactions" (Burns and Worsley 2015, p. 30). The key is to focus on hearing and amplifying each voice in community.

Different to implementing one project, complexity theory helps us see the larger picture. It explains a situation where people act "in a dynamic field where things are constantly happening" (Burns and Worsley 2015, p. 169). It is not sufficient to focus on one piece alone. Further, one cannot separate planning from acting. Things shift constantly. Dialogue is important. However, "agreement is signaled by engagement" (Burns and Worsley 2015, p. 169). People adopt the practices that work for them, which creates ownership from the beginning (Burns and Worsley 2015, p. 169). What people choose to do and why is important to reflect on throughout.

In essence, the practitioner should be a facilitator and a mediator. The poorest hold the most knowledge about their own situations. The practitioner has knowledge but it is different, external, while the poorest have intimate knowledge of their own circumstances.

Facilitation and facilitators are key. A facilitator must be able to work with many different types of people. Facilitators need to: "Listen ... Learn from experience ... Analyze who has a stake in an issue ... Adapt to new situations. ... Admit uncertainty, mistakes ... Provide constructive criticism ... Build relationships. ... Mediate differences" (Cohen 2001, p. 30). Practitioners need to develop skills in communication, mediation, and openness, as well as be open to learning and having their own perspectives shifted. In particular, in community settings, practitioners need to know the language(s), local organizations, and regional and national contexts. Practitioners need to build long-term relations with the community and enhance and encourage local people's capabilities.

The facilitation and facilitator become part of the community dialogue. Victoria Fontan, scholar in Peace Studies, states that a dialogue is different to a discussion or conversation. In a discussion, Fontan explains, people tend to put forth their own arguments, holding fast against the other's perspective. So too with a conversation although less intensely. In contrast, "A dialogue revolves around the joint creation of new knowledge" (Fontan 2012, pp. 163–4). A dialogue is an opening for speaking, listening, and learning. Each perspective is valid and together these lead to new perspectives and knowledge.

The role of a facilitator then is to encourage and enable local communities to gather to "define their own issues, objectives, and strategies based on their needs and wants" (Cohen 2001, p, 12). As the community proceeds, the facilitator helps to "identify commonalities within groups and communities that may be divided by gender, race, class, and other differences" (Cohen 2001, p. 12). The facilitator supports the group by amplifying the voices beyond the group and enabling learning (Cohen 2001, pp. 12–13). Rather than lead a project, the practitioner enables the community to define its goals and then achieve those goals, mediating differences, and with constant reflection.

One starting point for community dialogue could be social mapping. As a group, the community maps out its households and other criteria it deems important. Often this is done on the ground so that many people can add things at once. In this way, the group learns how they each see things, building a fully picture of the whole.

The facilitator should include each person in the community, as we all see our surroundings and situation differently. For example, "In a village in Sierra Leone where men and women mapped separately men showed interest in roads and junctions, while women were concerned with the well and hospital and wanted them close by" (Chambers 2008, p. 142). The facilitator's role is to ensure each voice is heard. As we explore in the chapter on intersectionality, there are many ways in which people are excluded. The goal here is to include every voice.

The mapping can also become issue based. In systemic issue mapping, the community creates a map of issues together. The aim of the process is to "develop insight into problems, why they emerge, how they are maintained, and how they become entrenched" (Burns and Worsley 2015, p. 65). The group tells, collects, and analyzes stories, then forms maps to try to show factors and connections. Once the map is drawn, the community analyzes it, looking for issues, connections, patterns, power relationships, and so forth. Then the map is distilled into one or more smaller maps, each examining a system, sometimes with smaller groups of participants. Then, the distilled maps are brought back to the larger group for checking (Burns and Worsley 2015, p. 79).

One powerful tool emerging from this mapping is called counter mapping. Here community maps are used as evidence to defend the perspective of the marginalized (Chambers 2008, p. 43). The maps generated by the community often fill knowledge gaps at regional and national levels or are different to those understandings.

GIS (Geographical Information Systems) technology has been made available to many communities, a good example of sharing technology that we need more of. GIS is mapping software with the ability to analyze and reflect on the data. It can also help coordination of disparate organizations. The data is created by the communities themselves and remains available to the communities for updating and analysis. Communities learn how to use the program, which offers "greater spatial accuracy, permanence, authority and credibility with officialdom" (Chambers 2008, p. 138). GIS mapping has helped with: "protecting ancestral lands and resource rights; management and resolution of conflicts over natural resources; collaborative resource-use planning and management... [and so forth]" (Chambers 2008, p. 139). Communities have used participatory GIS in Brazil, Cameroon, Canada, Ethiopia, Fiji, Ghana, just to name a few countries. My dream would be for it to expand in Haiti, in particular, as the situation between local communities, NGOs, and local, regional, and national government is so complex and fragmented.

Beyond mapping, communities are also capable of assessing progress and change. Another arena is participatory statistics. Jeremy Holland, social development consultant, offers examples of participatory mapping, modeling, proportional piling, card writing, sorting, ordering, and positioning, matrix ranking and scoring, among other methods (Holland 2013, p. 3). Using community ways of measurement, clear understandings of positive and negative change can be articulated. Participatory statistics can become a way for the community to hold NGOs and governments accountable (downward accountability). The poor should decide whether the help is working.

One example is found in Ashish Shah's chapter on participatory statistics in Rwanda for community planning. Rwanda has a tradition of Ubudehe, which is "collective action, mutual help, and reciprocity to solve community problems" (Shah 2013, p. 49). For more than a decade, Rwanda has encouraged and formalized this process. Ubudehe has generated a national census, among other things, which helps to inform the national government about local areas. "To date this has included district-level use of Ubudehe data for targeting and prioritizing district investments and for holding district officials accountable for outcomes under their performance contracts" (Shah 2013, p. 49). The outcomes are also local and regional, strengthening local participation in the political realm.

Participation in the political arena is crucial. Participatory democracy can be encouraged. Some emerging methods include: "citizen's juries (Wakeford et al., 2008), participatory budgeting, budget tracking, monitoring of service delivery, responsibility mapping, report cards, [etc.]" (Chambers 2008, p. 170). Communities can hold local, regional, and national governments accountable. Where NGOs may want to focus on governance from the top-down, what is critical is ensuring governments listen to the voices of the local communities, in particular, the poorest. Every government in the world should listen to its poorest.

While the community should ultimately decide how to proceed, these starting points could be suggested as they have proven useful in many settings.

How to mediate between the North and the South

The development practitioner also needs to advocate for change in the North based on the demands of the poorest. It is not enough to reduce poverty, while leaving the wealthy and powerful with their wealth and power. NGOs need to (1) be with the poorest, (2) work with the poorest, and (3) advocate for change in the global North.

There can be a tension between conscientization and advocacy. With conscientization, people gain the power to change their own lives. Advocacy to change the system, in contrast, comes from the privileged, often acting on behalf of the poor. While people with power may use their power for good, there remains an unequal distribution of power. Both empowerment and advocacy are critical. However, the advocacy must amplify the voices of the powerless.

The barriers that keep people powerless originate in the global North and among the privileged across the globe. The voices of the poor need to be brought to the North.

Development discourse convinces us that what "poor people" need is technical assistance, humanitarian aid, and resources. That assumption obscures the reality that people are poor because political-economic structures make them poor while making others rich. In this sense, *development contributes to the moral and political oblivion of economic privileged people of the global North.*

(Moe-Lobeda with Helmiere 2013, p. 212)

Just as NGOs can enable governments and economies to continue ignoring the poor, NGOs can also enable the global North to ignore its own need for change. When an NGO focuses on asking for donations, particularly for emergencies, it reinforces the assumption that the poor need the help of the rich. Requesting donations is important but it should not be the main focus. There needs to be a sustained strategy of educating the public on major issues. This happens to a certain extent with campaigns but again, campaigns tend to choose one issue at a time. We need to think more in terms of education than sound bites.

You can think about the development practitioner as a mediator between the global North and South between people who consider themselves developed and those experiencing development. The practitioner to be able to facilitate groups of the poorest and articulate the situation to groups of the powerful. One needs to be able to move between cultures and help those cultures learn from each other. For change to occur in the North, the North needs to know how it harms others.

Chambers suggests treating elites as potential allies, regardless of their initial position. Practitioners can provide elites with information they can use to speak with others. Within organizations, one can look through official documents to see where points of agreement are found. In particular, practitioners need to offer information from those experiencing development themselves (Chambers 2012, p. 158). Farmer adds,

> If we assume that we might be able to make headway on these problems by working ... with other people who may be policy-makers, by giving them information and the chance to play a constructive role and to undertake long-term ...we have to do that.
>
> (Groody 2013, p. 169)

Rather than assuming the powerful are the enemy, we can assume the powerful will be on board once they understand the situation. This perspective works as an attractor, according to complexity theory. Rather than presenting information as if someone will disagree, present the information as if there will be agreement and work from there. For Christian organizations, articulating the theology offered here can be useful.

Concluding suggestions

FBDO action:

1 Be with the poorest. Walk with the poorest.
2 Amplify the demands of the poorest to the powerful.
3 Articulate how to change structures to remove the barriers that keep people powerless.

Because the process of realizing the goals of the most marginalized is complex, as the chapter on intersectionality will address, working with the marginalized must be flexible, adaptive, and long term. Donors and practitioners need to commit to communities for the long term, responding to the local community's assessment of their own needs.

Note

1 LSE does currently offer "Cutting Edge Issues in Development Thinking and Practice but this course is unassessed. www.lse.ac.uk/study-at-lse/Graduate/Degree-programmes-2019/MSc-Development-Studies. IDS, Sussex, continues to be an exception to the rule, with "Economic Perspectives on Development" as an option but not a requirement for the degree. SIT is another exception with an interdisciplinary program including fieldwork, as is Concordia University, Portland. Development Studies programs do however focus on improving the situation of the poor rather than reducing the power and wealth of the rich and do not tend to include core courses on either religion or ethics.

References

Alghaib, Ola Abu. 2015. Aid Bureaucracy and Support for Disabled People's Organizations: A Fairy Tale of Self-Determination and Self-Advocacy. In *The Politics of Evidence and Results in International Development: Playing the Game to Change the Rules?*, eds. Rosalind Eyben, Irene Guijt, Chris Roche, and Cathy Shutt, 115–33. Bourton on Dunsmore: Practical Action. doi:10.3362/9781780448855.007

Balboa, Cristina. 2018. *The Paradox of Scale: How NGOs Build, Maintain, and Lose Authority in Environmental Governance.* Cambridge: MIT Press. doi:10.7551/mitpress/11254.001.0001

Burns, Danny and Stuart Worsley. 2015. *Navigating Complexity in International Development: Facilitating Sustainable Change at Scale.* Bourton on Dusmore: Practical Action. doi:10.3362/9781780448510

Chambers, Robert. 1988. Bureaucratic Reversals and Local Diversity, *IDS Bulletin*, vol. 19.4, 50–6. doi:10.1111/j.1759-5436.1988.mp19004008.x

———. 1993. *Challenging the Professions: Frontiers for Rural Development.* London: Intermediate Technology Publications. doi:10.3362/9781780441801

———. 1997. *Whose Reality Counts? Putting the First Last.* London: Intermediate Technology Publications. doi:10.3362/9781780440453

———. 2008. *Revolutions in Development Inquiry.* Abingdon: Earthscan. doi:10.4324/9781849772426

———. 2012. *Provocations for Development.* Bourton on Dunsmore: Practical Action. doi:10.3362/9781780447247

———. 2017. *Can We Know Better? Reflections for Development.* Bourton on Dunsmore: Practical Action. doi:10.3362/9781780449449

Cohen, David. 2001. Lessons from Social Movement Advocacy. In *Advocacy for Social Justice: A Global Action and Reflection Guide*, eds. David Cohen, Rosa de la Vega and Gabrielle Watson, pp. 11–31. Bloomfield: Kumarian Press.

Fontan, Victoria. 2012. *Decolonizing Peace.* Lake Oswego: Dignity Press.

Freire, Paulo. 1970. *Pedagogy of the Oppressed.* Trans. Myra Bergman Ramos. New York: Herder and Herder.

———. 1974. *Education for Critical Consciousness.* London: Sheed and Ward.

———. 1975. *Cultural Action for Freedom.* Original Edition 1970. Harmondsworth: Penguin Education.

———. 1993. *Pedagogy of the City.* New York: Continuum.

Groody, Daniel, interviewer. 2013. Reimagining Accompaniment: An Interview with Paul Farmer and Gustavo Gutiérrez. In *In the Company of the Poor: Conversations with Dr. Paul Farmer and Fr. Gustavo Gutiérrez*, eds. Michael Griffin and Jennie Weiss Block, 161–88. Maryknoll: Orbis Books.

Haynes, Jeffrey. 2007. *Religion and Development: Conflict or Cooperation?* New York: Palgrave. doi:10.1057/9780230589568

Hinton, Rachel and Leslie Groves. 2004. The Complexity of Inclusive Aid. In *Inclusive Aid: Changing Power and Relationships in International Development,* eds. Leslie Groves and Rachel Hinton, 3–20. London: Earthscan.

Holland, Jeremy. 2013. Introduction: Participatory Statistics: A 'Win-Win' for International Development. In *Who Counts? The Power of Participatory Statistics*, ed. Jeremy Holland, 1–20. Bourton on Dunsmore: Practical Action. doi:10.3362/9781780447711.001

Lwijis, Janil. 2012. NGOs: What Government Are You? In *Tectonic Shifts: Haiti Since the Earthquake,* eds. Mark Schuller and Pablo Morales, 69–72. Sterling: Kumarian Press.

Moe-Lobeda, Cynthia with Frederica Helmiere. 2013. Moral Power at the Religion-Development-Environment Nexus. In *Handbook of Research on Development and Religion,* ed. Matthew Clarke, 201–19. Northampton: Edward Elgar. doi:10.4337/9780857933560.00019

Reifenberg, Steve. 2013. Afterword. In *In the Company of the Poor: Conversations with Dr. Paul Farmer and Fr. Gustavo Gutiérrez,* eds. Michael Griffin and Jennie Weiss Block, 189–97. Maryknoll: Orbis Books.

Ritzer, George. 2009. *Globalization: A Basic Text.* Chichester: Wiley-Blackwell.

Shah, Ashish. 2013. Participatory Statistics, Local Decision-Making, and National Policy Design: *Ubudehe* Community Planning in Rwanda. In *Who Counts? The Power of Participatory Statistics,* ed. Jeremy Holland, 49–63. Bourton on Dunsmore: Practical Action. doi:10.3362/9781780447711.004

Shutt, Cathy. 2015. The Politics and Practice of Value for Money. In *The Politics of Evidence and Results in International Development: Playing the Game to Change the Rules?*, eds. Rosalind Eyben, Irene Guijt, Chris Roche, and Cathy Shutt, 57–77. Bourton on Dunsmore: Practical Action. doi:10.3362/9781780448855s

5 Justice in practice
Economics

For much of its history, development has had a double aim: economic growth with poverty reduction. Yet, the two aims have proven contradictory. I argue that within development, economic growth should not be a focus. An economy encompasses far more than growing money. Economics at its most basic is the study of how to distribute resources. Further, poverty is far more expansive than economic poverty. FBDOs can play an important role in shifting the economic conversation and expanding the conversation beyond economics.

This chapter presents some economic history, major foci of capitalism today, and suggests alternative economic foci. The following chapter moves beyond economics to argue for a holistic approach to development. Beginning with an analysis of the economic system is important because in our current global situation, the economic system wields far more power than governments and civil society. In terms of economics, poverty and wealth need to be eradicated achieving an economic situation where people are empowered rather than dominated.

More than half of the 100 largest economic units in the world are companies, not countries. The nine richest people in the world are worth more than 600 billion dollars, more than the poorest half of the world: 4 billion people combined (Jacobs 2018). The assumptions and practices of capitalism have not been dominant through history. Other economic systems existed and continue on a small-scale today. Capitalism has congealed power in parts of the global North. Development practitioners need to articulate this harm and how alternatives can shift the structures.

A brief economic history

A variety of economic systems have existed throughout human history, reciprocity, distribution, and householding being predominant until capitalism emerged. With hunter-gatherers (foragers), for example, there was a culture of gift-giving, reciprocity. These groupings of less than 100 people were predominant until 11,000 BCE though some still exist today. With the move toward pastoralism and agriculture, food became a commodity. "Once food becomes a commodity, criteria are simultaneously developed that determine who merits food and who does not" (Gustafson 2015, p. 115). Here, reciprocity was more measured. The beginnings of accumulation and inequality emerged.

The concept of money emerged around 3500 BCE, first as an accounting mechanism. In Mesopotamia, records were kept of what was owed to the center by the outer regions (Gustafson 2015, pp. 117–20). The first forms of private property emerged much later around the late 8th century BCE in Greece and the Near East (Duchrow and Hinkelammert 2004, p. 5).

The understanding of money shifted around 600 BCE as coins were created. Instead of simply keeping accounts, and trading goods or services, one could now use coin as payment. Soon after, the concept of owning property also narrowed, at least in the Roman Empire. "(*Dominium* or *proprietas*) ... designates a comprehensive right to a thing, a 'full right' not limited in time" (Duchrow and Hinkelammert 2004, pp. 12–13). However, other cultures defined ownership differently. For example, for the Israelites at this time, as we saw in previous chapters, God owned the earth. Humans were to use the land for the good of all.

Coins ceased to exist again in the European context around 600 CE, with money limited to accounting. Under feudalism in Europe, every social class had its own particular duties (Gustafson 2015, p. 126). During the time span of European feudalism, other regions around the world were more "advanced" in terms of empire and trade. Arab Islamic empires emerged and expanded in the Middle East after 600 CE and into Africa, slowly shrinking after 1100. By 1350, the Chinese had an advanced political and economic structure, with several scientific achievements, including shipbuilding, sailing, and soon after gunpowder. Around this time, the Ottoman empire emerged and expanded. In the "undiscovered" Americas, numerous groups of indigenous existed, including the hierarchical civilizations of Inca, Maya, and Aztec, among others.

Coins reemerged in Europe in the late 1400s.[1] Capitalism began to emerge in a few different places in Europe while at the same time Europe benefitted from the scientific knowledge produced by the Chinese and the Muslim world. For example, in the 16th century, citizens of Genoa began to travel to sell their goods, using bills of exchange for long distances. At the same time, Spain invaded South America and the

Caribbean. Portugal aimed itself toward Asia and Africa. The Dutch emerged as a power with the "triangle trade." Slaves were kidnapped from African regions and sold in the Americas, where they were forced to labor for the colonists. The materials produced were shipped to Europe, where laborers manufactured and owners sold the goods.

England emerged as a power with money as a paper note beginning in 1694. Banks could take deposits and give money back as paper. Banks could also loan paper money. With a loan, you receive money and the bank also considers itself to have that money, as they expect you will pay it back. So the sum of money doubles (Gustafson 2015, pp. 128–9). Money could now create more money.

England and others industrialized production, leaving the poor with less access to work and land. The USA fixed its dollar to the gold standard in 1900. After World War II (WWII), other countries fixed their exchange rates to the dollar. At the same time, governments began to address the increasing poverty in Europe and the USA, due to the World Wars and the 1929 stock market crash, leading to varying levels of welfare states.

Development emerged and consolidated after WWII and during the Cold War. Regions of Asia and Africa were gaining independence from former colonial powers; Latin American countries had gained independence throughout the 1800s. The goal was to modernize poor countries to help them reach the level of the USA and Europe. Likened to an airplane's journey, the stages included: "traditional": the parked airplane, "pre-take-off," "take-off," and the "drive to maturity": the flight, achieving "mass consumption" at the end of the trip (Clarke 2013, p. 3). As the Cold War became a focus, democratically elected Latin American governments were taken over by military coups, arguing that governments needed to protect these countries from communism (with U.S. support). The same occurred in the newly independent nations of Africa and in Asia. While development was supposed to include capitalism and democracy, it was experienced as capitalism and dictatorship. Aid for development focused on Northern country goals.

The large-scale projects did not improve the lives of the poorest. In Latin America, dependency theory emerged, arguing that richer countries caused underdevelopment by their own development. Rather than a linear progression, development caused a widening gap. Postcolonialism also emerged, critiquing colonialism and its lingering effects.

In 1971, U.S. President Nixon ended the linking of the dollar to gold. Money was now free-floating, subject to speculation on its value. Further, oil prices tripled causing expenses for oil-poor nations and wealth for oil-rich nations. These "petrodollars" went to banks that then needed to lend the money out, to generate more money. The World Bank expanded its focus from large-scale projects to meeting "basic" needs. Poverty would be eradicated by loans to generate growth. The two aims were incompatible.

The loans led to huge debts and capital flight: "*Dubious debt*" refers to loans that dictatorial governments took out and used for huge projects that did not benefit the poor or were used for different purposes. Some loans went straight into the pockets of elites, who then deposited this money back in the global North (Henry 2006, p. 253). Together, this wealth is estimated to be nearly $9 trillion. The misuse of loans generated more wealth for the North too.

By the time the global North supported the move back to democratic governments in the 1980s and 1990s, the loan crisis had deepened. Steven Hiatt, U.S. writer and activist, states that the global South pays $375 billion annually on its loans and receives less than 19 billion per year in aid (Hiatt 2006, p. 13). The World Bank and the IMF offered further loans, so countries could keep paying back the previous loans. (A country cannot declare bankruptcy.) The conditions attached came to be known as structural adjustment, neoliberalism, and the Washington Consensus. Conditions included ending government provision of basic services and opening markets.

The focus shifted to globalization and trade with the forced opening of markets and the shrinking of government services. Some companies became more powerful than many countries and people started agitating against the unpayable debts. Multinational corporations tend to be based in richer countries, export one-third of world goods, and make up 10% of global GNP (Ritzer 2009, pp. 198–9).

Development appeared to be in crisis; it was stagnating. Post-development emerged in the 1990s, rejecting the neutrality of the concept of development. Post-development coalesces around the idea that the overconsumption of the global North is the wrong goal.

Toward the end of the 1990s, the World Bank's approach began to shift from structural adjustment toward "capabilities," as articulated by Amartya Sen, Indian economist and philosopher.[2] Having mostly finished the desired economic opening of markets, and with campaigns, as well as clear evidence that some governments could not continue paying debts, some debt relief occurred. Millennium Development Goals were created and now we have Sustainable Development Goals. Groves argues the focus shifted again from local communities back to governments (Groves 2004, p. 77). While bypassing governments is not a good idea, the loss of focus on local communities is problematic. Many countries, including the USA, need to empower local communities to engage regional and national governments. See for example, the failure of the U.S. government to respond appropriately after Hurricane Katrina in the southern USA (Adams 2013).

Having affected poor country economies, the World Bank and IMF began to require attention to "governance" in order to receive loans, "to foster improvements in the way in which public institutions conducted public affairs and managed public resources" (Burns and Worsley 2015, p. 13). Known as the post-Washington Consensus, the aim is to encourage the work of NGOs and to improve governments.

Aid was reduced further after the U.S. and European financial crisis of 2008. However, the powerful countries ensured their own recovery. Wealthy country governments bailed out their banks and large corporations. Average citizens felt the hit as corporations recovered and continued to profit.

Today, a very small portion of the population owns the majority of resources and money. Further, the economy is far more complicated than producers and consumers. One analysis of tropical products, centered on coffee, listed the following chain: "producer, primary processer/middle person, exporter, international trader, industrial processor, wholesaler, retailer and consumer" (Daviron and Ponte 2005, p. 2). And this list is only for an actual good. The chain does not include the realm of finance and service in general. There are many complicated layers to capitalism.

The process of globalization broadly aims toward a global economic system: capitalism and a global political system: democracy. Globalization's effects have been uneven, enriching some and impoverishing others. As Kim argues, "Globalization may be viewed as the creation of a single, international financial order which has left most of the poorer countries buried under huge debts. ... *The new natural resource of poor countries is poor, unregulated, and unprotected labor*" (Kim 2013, p. 13). With the forced opening of markets, companies shift production to countries with the lowest wages. There are contested versions of what and how to globalize and people experience vastly different effects. There is no one simple definition of globalization.

Underlying assumptions of capitalism

Let us first return to the problems with capitalist assumptions and inherent in capitalism itself. I focus on capitalism here because it is the economic system currently being globalized. Critiques can and have been made about other economic systems as well. New alternatives need to be created as I articulate later in the chapter.

First, capitalism focuses on the market, as the conduit of exchange between people. If one cannot participate in the market, one does not count. A second assumption is that we act in the market based on rational self-interest. In theory, each person enters into the marketplace freely. In practice, we enter the market at varying stages of life with varying obligations. Further, I may also only have access to certain items, for example, I might live in a food desert, or I may have an urgent need such as health care costs that outweigh longer-term rational planning.

Third, because capitalism focuses on wealth accumulation, it has no clear definition of poverty or wealth. Economists focus on wants rather than needs, despite the fact that the majority of the world's population is still aiming for basic subsistence. Economists focus on the minority with access to greater sums of money. Economists also do not address an upper limit of accumulation. We can also see through advertising that people do not follow "unlimited consumption" without urging. Advertising teaches people that they should want a particular product. Companies find they have to tell consumers what to desire and make it difficult for consumers to sustain what they already have, in order to actually create desire.

Fourth is the assumption of scarcity based on humanity's unlimited desires. Money and natural resources are both scarce in this understanding. Money in our society is simply a piece of paper or a few words on a computer. Yet, a person can do little without these words or papers. The less money you have, the less access you have to credit. Competition emerges here for the scarce resources.

Two conflicts emerge here. First, with scarcity, growth is limited. Second, competitions end with one winner, or what we call a monopoly. With competition, more and more people or companies drop out of the contest as one company accumulates more or a person receives that particular job or purchases that house.

Labor is considered last, if at all in capitalism, although it is impossible for any product to exist without some form of labor. Claar and Klay state that "economists... are almost uniformly opposed to any law that would artificially determine wage rates apart from market forces" (Claar and Klay 2007, p. 172). Since the market should be left to its own devices, setting a minimum wage would be interference. Claar and Klay argue that setting a minimum wage harms the poorest who cannot find a job because fewer jobs exist since employers cannot pay that minimum. There are "people who are so disadvantaged... that they are dying for just a chance to work at $5 per hour" (Claar and Klay 2007, p. 175). However, they also state that "among the resources that are in scarce supply, and therefore require markets to promote efficient utilization, is labor" (Claar and Klay 2007, p. 168). If true, there would be no unemployment: unused labor. No one would be "dying" for a chance to work at $5 per hour.

Beyond these underlying assumptions, capitalism contains inherent problems. First, rather than seeing extreme poverty and inequality as a problem caused by capitalism, capitalists see it as a lack of completely implementing capitalism and as necessary to economic growth. "Freedom always means change- and change creates some losers (especially in the short run) as well as winners" (Claar and Klay 2007, p. 36). Some people will be left behind.

Second, economic growth is measured at a country level through the Gross Domestic Product (GDP). Growth comes from profit and profit from an increase in money; all this is considered good. However, events like car accidents, and injuries all add to GDP, through any repairs, insurance claims, doctor and hospital visits, etc. Further, growth ignores how the wealth of a country is distributed. It would see no difference between one person with a billion dollars and 999 having 0 or each of a thousand people with one million dollars.

Third, in order for the economy to grow, there must be debt and credit. New money must be created.

> If they [loans] are repaid at interest, then more money must be eventually withdrawn from the economy than was originally added. ...[Money] must be replaced by the creation of further money elsewhere in the form of a loan. The entire economic system functions as a spiral of increasing debt ... Debt will then need to be redeemed by massive levels of inflation, devaluation and impoverishment.
>
> (Goodchild 2009, pp. 69–70)

New money is created through loans, causing exponential debt and then recession. So far, these economic collapses have mostly affected the already poor and marginalized, due to government intervention, as in 2008.

Fourth, capitalism calculates the use of the earth's resources as an addition to economic value rather than a subtraction. For example, the use of oil removes that oil from the earth. It is non-renewable (at least for millions of years). Hence, a resource is reduced but the economy considers this an addition. The ecological damage of these calculations has also been severe. Two aspects are "climate imperialism," which explains how wealthier nations contribute the most toward climate change but poorer countries suffer the worst effects, and "environmental racism," where richer countries deliberately send waste, toxins, and polluting companies to poorer countries and poorer communities within the wealthier countries (Moe-Lobeda with Helmiere 2013, p. 204).

Fifth, the market treats life forms as commodities that can be owned by one person. The US Supreme Court, for example, has allowed plants and DNA to be patented, leading to severe consequences. Some multinational corporations have patented seed and then sell seed whose plants do not produce seeds, so farmers have to buy seed every year and from one company. Corporations are also patenting many of the genes that make up human DNA.

Sixth, justice is limited to its commutative form. In terms of justice, they state: "we practice justice by giving all persons their due as humans, made in God's image. In labor markets this includes paying people their agreed-upon wages, based on the value they contribute to production" (Claar and Klay 2007, pp. 218–9). Distributive and social justice are absent here, although governments do sometimes try to encourage them.

However, capitalists prefer a government to support capitalism, rather than curtail its effects. Capitalists argue the role of government is to keep the market flowing freely with little to no intervention. Governments are considered problematic for three reasons: First, "all citizens are forced to participate" (Claar and Klay 2007, p. 215). Yet, this is a space where people could participate without money. Frankly, people are forced to participate in the market too, if they can access money. Second, "governments sometimes mismanage public resources" (Claar and Klay 2007, p. 216), a statement which ignores the role of active citizenship. Third, "when a government undertakes to provide a public service, ... the result can be less total help, because private individuals may assume that their moral duty to help those in need is adequately covered by their taxes" (Claar and Klay 2007, p. 216). Again, this argument retains a focus on charitable giving rather than redistribution toward justice.

Moving toward justice

Let us consider what alternative economies that prioritize people could look like, remembering that the poorest need to decide for themselves. To begin to regain the balance between realms, certain ethical decisions could be taken: "to confirm that human life has a greater value than that of the market" (Houtart and Polet 2001, p. vii). An economy could prioritize people over money.

In particular, within the economic system, we should focus on the poorest and help to change the systems that keep people poor. Meeting the basic needs of the poor, as defined by the poor, would be our first focus. This focus includes advocating for structural changes. Then we could focus on preventing poverty.

One could begin with a Citizen's Income, which would provide a basic amount of money to each person (Hines 2000, pp. 246–7). In poorer countries, this income could be achieved by cancellation of debt and larger amounts of aid until we move toward a more equal footing. Further, goods necessary for survival, such as water, should be freely accessible.

In considering people at the center, work in order to sustain one's family and community becomes important. Reconnecting people to actual products of work and meaningful work is key. Capitalism encourages "efficient" production, which means breaking down the task into its component parts and having machines or humans work on one part or piece, as a specialty. Creativity is limited to the one who has the idea and enough money to enact it. To counteract this disconnect, the concept of economic democracy can be useful.

Economic democracy means economic decisions are made at local levels, with full participation, for example, worker-owned cooperatives or businesses. Hans Binswanger, Swiss economist, as cited in Duchrow and Hinkelammert, argues for: "*personal development* and the *economic safeguarding of all* those working in the company," enabling each worker to participate in company decisions, and worker-owned businesses (Duchrow and Hinkelammert 2004, pp. 178–9). Workers need to have a say in their work situations, ideally part-owning the business. Investors should remain actively involved with the businesses in which they invest. Speaking of economic democracy in the Caribbean context, Erskine argues, "Space should be created in which the poor are allowed to participate in the creation of institutions that have their best interests at heart" (Erskine 2009, p. 283). Institutions should not be created for the poor, they should be created with the poor.

Alternatives to the current economic system include, for example, participatory economy: Parecon. "Parecon seeks to fulfill four key values: solidarity, diversity, equity, and self-management" (Albert 2005, p. 3). Institutions that will help achieve this include: "worker and consumer councils" (Albert 2005, p. 9). Further, a sustainable economy would transparently include values. Goodchild argues that "evaluative credits should circulate alongside money, goods and services" (Goodchild 2009, p. 246).[3] In order to ensure values are a part of the economy, Goodchild proposes evaluation alongside money.

In economic democracy, one begins locally, then regionally, then nationally, then internationally. In aiming for sustainable local economies, (1) what can be produced locally should be protected from imported goods, (2) businesses should be located where they sell goods, (3) local money can be created (LETs), (4) taxes should be progressive, (5) economics and politics should begin at the local level, and (6) the poorest communities will be enabled to achieve these goals (Hines 2000, p. viii). Here, the emphasis is on addressing the needs of the poorest and restoring ecological resources, moving to the use of renewable resources. One way to achieve this sustainability is called bioregionalism. Another key suggestion from ecological economics is that any good that is produced needs to include in its pricing the cost for the environmental resources, potential waste, and the disposal of a product, in addition to the cost of production.[4]

It is not enough to raise the wealth of the poorest. We need to reduce the wealth of the richest too, shifting the focus of economies to sustaining people, rather than profit and growth. For example, the world's nine richest people's combined worth is more than 600 billion dollars, greater than that of half of the population of the world (Jacobs, 2018). Amazon, led by Jeff Bezos, the world's wealthiest man, has not paid tax for the previous two years, despite huge profits. Leaving each of the nine with a net worth of one billion would still free up 600 billion dollars. Redistributing that wealth to the poorest half would come to roughly $150 per person, which is significant given that many people live on less than $2 per day. Such an asset would improve the income of the poorest, negatively affecting nine people. If done once a year for five years, it would negatively affect 45 people and redistribute a minimum of 1,600 billion dollars of wealth. It would also incentivize the wealthiest to give away more so as not to end up in the top nine.

Simple cash transfers (CTs) can help the poorest, like the very beginnings of a basic income. For example, the Kenya Cash Transfer for Orphans and Vulnerable Children has reached 25,000 households and the initial results are positive (Alviar, Ayala, and Handa 2010, p. 97). Another study of an unconditional CT project in Vietnam organized by Oxfam stated

> the analysis demonstrates how the poor can use money responsibly when given the chance; how this money can help the poor address some structural aspects of their poverty and finally how it can lead to social transformation in the focus villages
>
> (Chaudhry 2010, p. 169)

Giving small amount of cash enables the poorest to begin making decisions.

Broadly, the state's role should to be protect and sustain its citizens and to empower its citizens to decide how to accomplish this role. We will discuss the limits of the nation-state in the following chapter.

Action

National

National governments need to make several changes to focus on the poorest: implement progressive taxes, reduce spending on defense, and protect local businesses and resources. Welfare systems should cover basic needs, including health care and education. Water, transportation, communication systems, and renewable energy must be publicly available. For example, the U.S. economy could shift its focus to public transport, single-payer health care, shifting food supply costs and taxes to make health foods more affordable than unhealthy foods, renewable energy, and connecting home mortgages to green efficiency (Connolly 2008, pp. 105–8). In regulating business and economies, governments could encourage economic democracy, a focus on the local, including LETs and cooperative banking, full cost accounting, sharing of technology, and fair trade.

International

Beyond national action, there should be international action. In the international realm, using loans as aid should end. Money should be given not lent. As for past loans, international debts from any developing country that has already paid more than the principal borrowed must be cancelled. Further, much is owed to the South by the North. Ilo, citing the African Climate Justice Manifesto, argues:

> The developed countries ... must repay their debt through deep domestic emission reductions and by transferring the technology and finance required to enable Africa to follow a less polluting pathway. ...[Wealthy countries] must compensate Africa for the adverse effect of their excessive historical and current per-person emissions.
>
> (Ilo 2011, pp. 275–6)

Wealthy countries also need to atone for their overuse of resources.

Currency regulation should be reintroduced and speculative finance ended. Tax flight and avoidance must be ended. Prior tax flight and avoidance by companies and politicians must be redressed. In particular, poorer countries should receive help to recover the wealth elites siphoned off.

New international agencies could help to achieve these goals, for example, "an International Asset Agency, a Global Investment Assistance Agency, and a World Trade Agency… [which] would work to attain equity, solidarity, diversity, self-management, and ecological balance in international financial exchange, investment, development, trade, and cultural exchange" (Albert 2005, pp. 58–60). International institutions should protect the poorest, not the wealthy.

Church

Churches can also play an important role in these changes. Churches need to support local and cooperative ownership, advocate the ending of patenting of life forms, ensure that environmental goods are cooperatively or publicly owned, ensure that private property is used for the common good, support progressive taxation, advocate to end speculative finance, tax flight and avoidance, and advocate for democratic international institutions (Duchrow and Hinkelammert 2004, pp. 220–1). Taylor adds from the perspective of the poor that the churches and related organizations should: "draw attention to excessive wealth and greed as well as poverty" (Taylor 2003, p. 70).[5] Churches need to help educate followers on the concept of "enough."

Beyond broader advocacy, churches should ensure their own money is invested ethnically and that their land is used for the common good. For example, the largest landowners in the Caribbean are churches. "The membership, who are predominantly poor people, should insist on land reform, which should include the church making its vast resources of land available" (Erskine 2009, p. 283). The church cannot be prophetic when it benefits from the current economic system.

FBDO

1   Be present in the poorest communities.

    a   let the poorest decide how to improve their economies.
    b   at home and abroad, respond to those requests and advocate for change more broadly, based on those requests.
    c   help to create local economic councils.

2   Advocate for change in wealthier communities.

   a   encourage people to stop consuming from multinationals
   b   encourage people to buy local, organic, and fair-trade goods
       from employee owned and cooperative businesses. Some ex-
       amples of social responsible and/or worker owned companies
       include: Mondragon, the Mararikulam experiment in Kerala,
       India, Bruderhof communities in the USA, Canada, and
       England, John Lewis, Hyvee, Publix Super Markets, Scheels,
       Patagonia, Ocean Spray, Blue Diamond, Equal Exchange,
       Bob's Red Mill, Ace Hardware, and Kyocera.
   c   encourage people to use local banking and credit cards; en-
       courage banks to minimize the practice of interest on loans,
       and fees that impact the poorest.
   d   encourage people to invest ethically
   e   encourage people to advocate for these changes in their work-
       places and in the organizations with which people are in-
       volved (colleges, businesses, non-profits, churches, etc.).
   f   ensure your organization and associated churches/religious
       groupings do the above five things.
   g   help to create local economic councils.

3   Advocate for change nationally and internationally:

   a   encourage your government to stop subsidizing multinationals
   b   encourage your government to (1) tax any trading not di-
       rectly linked to production (Tobin tax, for example) with the
       goal to (2) end trading and investment not directly linked to
       production.
   c   encourage your government to focus trade policies on
       supporting local, organic, and fair-trade goods
   d   encourage your government to support local banking and
       credit and end subsidies to large banks and credit organi-
       zations unless they will solely invest ethically and minimize
       lending at interest.
   e   encourage your government to lobby for these changes in in-
       ternational organizations.

For everyone

Increase interactions between the rich and the poor; recognize how we
are interconnected. In these interactions, we need to treat each other
as equals, listening to the marginalized in particular.

Conclusion

Within development practice, enabling local communities to make their own economic decisions is crucial. Advocating these decisions regionally, nationally, and internationally can help to ensure local communities can survive and thrive. Development practice, however, must also engage the global North in reducing its economic privilege; it is not enough to attempt to improve the lives of the poorest without teaching the wealthy about "enough."

While economic alternatives are important to introduce, the new heaven and new earth will be much broader than the economic realm. By focusing on consumption and the marketplace, the global economic system excludes anyone who does not consume. While we do want to work to end exploitation, we also need to understand that there are people not even "privileged" enough to be exploited because the system excludes them completely. We can change the framework to move beyond economics. Local communities, as well as global ones, also suffer from inequalities in power due to race, sex, gender, religion, etc. The next chapter considers development practice at these intersections.

Notes

1 One major impact was Luther's revolt against Catholic practices, such as indulgences: a way to atone for sin. Indulgences had been action-based: prayer or service, etc. However, after coins re-emerged, people purchased the indulgence. Instead of God granting forgiveness and then the atonement, now atonement was purchased and then God granted forgiveness. (Gustafson, 126–7)

2 Amartya Sen developed a "capabilities approach" arguing that progress could be measured in terms of what people become capable of doing. While Sen has not developed a specific list, arguing that groups need to deliberate together on what to measure, Martha Nussbaum has articulated a list of "Central Human Capabilities," which include: "Life," "Bodily Health," "Bodily Integrity," "Senses, Imagination, and Thought," "Emotions," "Practical Reason," "Affiliation," "Other Species," "Play," and "Control Over One's Environment." (Nussbaum, 604–5)

3 "Each act of social evaluation may be regarded as a contract between four parties: 1. The one who makes the evaluation; 2. The one who receives the evaluation and carries out the investment …; 3. The institution that offers expertise in the making of an evaluation, termed an *evaluative institution*; 4. The institution that attributes social effectivity to such evaluations, termed a *bank of evaluative credit*." (Goodchild, 247)

4 Another way to distribute wealth comes from Thomas Pogge, a German philosopher, who suggests a global resources tax. Rather than shifting the economic calculation to count an extraction of a resource as a cost rather than a benefit, he suggests implementing a tax on any resource extracted

from the earth, emphasizing non-renewable resources. The tax collected would be used to reduce poverty.

5 Taylor offers a complementary "'2015 Millennium Goals' for the Churches: A Call to Action" (Taylor 2003, xiv), which includes: "3. Define a 'greed line' to stand alongside the 'poverty line' in each country, translating Gospel teaching on wealth into concrete and contemporary guidance for Christians. …5. Re-examine the reasons for supporting poverty-related projects and programs to make sure they are advocates of fundamental structural change in favour of the poorest." (Taylor 2003, xiv)

References

Adams, Vincanne. 2013. *Markets of Sorrow, Labors of Faith: New Orleans in the Wake of Katrina.* Durham: Duke University Press.

Albert, Michael. 2005. *Realizing Hope: Life beyond Capitalism.* New York: Zed Books.

Alviar, Carlos, Francisco Ayala and Sudhanshu Handa. 2010. Testing Combined Targeting Systems for Cash Transfer Programmes: The Case of the CT-OVC Programme in Kenya. In *What Works for the Poorest? Poverty Reduction Programmes for the World's Extreme Poor,* eds. David Lawson, David Hulme, Imran Matin, and Karen Moore, 97–114. Bourton on Dunsmore: Practical Action. doi:10.3362/9781780440439.006

Burns, Danny and Stuart Worsley. 2015. *Navigating Complexity in International Development: Facilitating Sustainable Change at Scale.* Bourton on Dusmore: Practical Action. doi:10.3362/9781780448510

Chaudhry, Peter. 2010. Unconditional Cash Transfers to the Very Poor in Central Viet Nam: Is It Enough to 'Just Give Them the Cash'? In *What Works for the Poorest? Poverty Reduction Programmes for the World's Extreme Poor,* eds. David Lawson, David Hulme, Imran Matin, and Karen Moore, 169–78. Bourton on Dunsmore: Practical Action. doi:10.3362/9781780440439.010

Claar, Victor and Robin Klay. 2007. *Economics in Christian Perspective: Theory, Policy and Life Choices.* Chicago: IVP Academic.

Clarke, Matthew. 2013. Understanding the Nexus between Religion and Development. In *Handbook of Research on Development and Religion,* ed. Matthew Clarke, 1–13. Northampton: Edward Elgar. doi:10.4337/9780857933577.00005

Connolly, William. 2008. *Capitalism and Christianity, American Style.* Durham: Duke University Press. doi:10.1215/9780822381235

Daviron, Benoit and Stefano Ponte. 2005. *The Coffee Paradox: Global Markets, Commodity Trade and the Elusive Promise of Development.* London: Zed Books.

Duchrow, Ulrich and Franz Hinkelammert. 2004. *Property for People, Not for Profit.* New York: Zed Books.

Erskine, Noel. 2009. Caribbean Issues: The Caribbean and African American Churches' Response. In *Religion and Poverty: Pan-African Perspectives,* ed. Peter Paris, 272–92. Durham: Duke University Press. doi:10.1215/9780822392309-014

Goodchild, Philip. 2009. *Theology of Money*. London: SCM Press. doi:10.1215/ 9780822392552

Groves, Leslie. 2004. Questioning, Learning and 'Cutting Edge' Agendas: Some Thoughts from Tanzania. In *Inclusive Aid: Changing Power and Relationships in International Development*, eds. Leslie Groves and Rachel Hinton, 76–86. London: Earthscan.

Gustafson, Scott. 2015. *At the Altar of Wall Street: The Rituals, Myths, Theologies, Sacraments, and Mission of the Religion Known as the Modern Global Economy*. Grand Rapids: Eerdmans.

Henry, James. 2006. The Mirage of Debt Relief. In *A Game as Old as Empire: The Secret World of Economic Hit Men and the Web of Global Corruption*, ed. Steven Hiatt, 219–61. San Francisco: Berrett-Koehler.

Hiatt, Steven. 2006. Global Empire: The Web of Control. In *A Game as Old as Empire: The Secret World of Economic Hit Men and the Web of Global Corruption*, ed. Steven Hiatt, 13–29. San Francisco: Berrett-Koehler.

Hines, Colin. 2000. *Localization: A Global Manifesto*. London: Earthscan.

Houtart, François and François Polet, eds. 2001. *The Other Davos: The Globalization of Resistance to the World Economic System*. London: Zed Books.

Ilo, Stan Chu. 2011. *The Church and Development in Africa: Aid and Development from the Perspective of Catholic Social Ethics*. Eugene: Pickwick.

Jacobs, Sarah. 2018. "Just Nine of the World's Richest Men Have More Combined Wealth than the Poorest 4 Billion People," *Independent*, January 17, 2018. Accessed August 12, 2019. www.independent.co.uk/news/world/riche stbillionairescombinedwealthjeffbezosbillgateswarrenbuffettmarkzuckerb ergcarlosslimwealth-a8163621.html

Kim, Grace Ji-Sun. 2013. *Colonialism, Han, and the Transformative Spirit*. New York: Palgrave Pivot. doi:10.1057/9781137344878

Moe-Lobeda, Cynthia with Frederica Helmiere. 2013. Moral Power at the Religion-Development-Environment Nexus. In *Handbook of Research on Development and Religion*, ed. Matthew Clarke, 201–19. Northampton: Edward Elgar. doi:10.4337/9780857933560.00019

Ritzer, George. 2009. *Globalization: A Basic Text*. Chichester: Wiley-Blackwell.

Taylor, Michael. 2003. *Christianity, Poverty and Wealth: The findings of 'Project 21'*. London: SPCK.

6   Justice at the intersections

Every human being is of equal value. Christians are called to focus on people treated as less than human. No one should be excluded from abundant life. Second, the marginalized should determine what development means for them. Not everyone wants the market at the center of life. FBDOs have an important role to play in addressing the complex issues that keep people at the margins. This chapter presents some of the issues emerging from the marginalized. This chapter argues for attention to intersectionality, all that marginalizes a person. Economic power intersects with race, nationality, sexuality, and so forth. We need to attend to this complexity.

The starting point is one's own place in reality and one's vision for the future. This simple starting point has often been dismissed within the development paradigm and globalization, which assume that people follow the same path. The problem with failing to acknowledge difference is that one concept of development does not make sense in every community. One's powerlessness is due to more than lack of money or access to the marketplace. One is even powerless to define development.

Development should focus on the most marginalized in community. Let us first ask: "who is excluded?" Then ask: "why?" Development focuses on the living, as it should, but we also need to ask, who has died prematurely and why? The dead cry out for attention. The prematurely dead have not been able to live life abundant and their deaths can help us to understand how to change.

First, we need to keep people alive. "We might not be undoing poverty by making sure we have high-quality medical care, but once we get this specific manifestation of poverty out of your body, that will leave you free to fight on against poverty" (Farmer qtd. in Groody 2013, p. 185). Providing sustainable health care is critical. Then, preventing illness is key. Farmer argues for public service provision, including health care, clean water, and sanitation. Such attention requires "contributing to stronger public health systems. ...It cannot be done, in other words, without changing the policies governing much, perhaps most, development assistance and humanitarian aid" (Farmer 2013a, p. 126). To prevent illness requires attention to multiple realms, the political in particular.

Many development projects attend to health, which is important. Health is critical to the poor: often their only asset is their body and its labor. Mander cites a *Lancet* study, which reported 5% of Indians fell into poverty because of health care-related expenses (Mander 2015, p. 114). Paying for health care can impoverish people. So free health care is crucial.

Living life abundantly goes further than not dying. It involves actually being able to live as a human being. People at the margins articulate impoverishment broadly, as the biblical texts do. Powerlessness includes being treated as an object and being ignored. One YWCA study found that "the women in that study described poverty as dehumanizing, all-embracing, and all-encompassing, and as 'an indivisible whole, an ongoing, day-to-day reality that cannot be simply defined by lack of a particular possession or amenity.'" (Bailey 2009, p. 53) Poverty includes lack of resources and lack of care for one's humanity.

However, Chambers argues that rather than asking what poverty is, the poor ask, "What can you do to reduce our bad experiences of life and living, and to enable us to achieve more of the good things in life to which we aspire?" (Chambers 2012, pp. 41–2) Rather than parse out a definition of poverty, we can enter communities and support them in their own efforts.

NGOs and others have expanded their definition of development beyond economic growth. However, NGOs and donors continue to determine the definition for the poorest, rather than letting the poorest define "good change." The Sustainable Development Goals (SDGs) do expand beyond economics, addressing 17 issues:

1. No Poverty; 2. Zero Hunger; 3. Good Health and Well-being; 4. Quality Education; 5. Gender Equality; 6. Clean Water and Sanitation; 7. Affordable and Clean Energy; 8. Decent Work and Economic Growth; 9. Industry, Innovation and Infrastructure; 10. Reduced Inequality; 11. Sustainable Cities and Communities; 12. Responsible Consumption and Production; 13. Climate Action; 14. Life Below Water; 15. Life on Land; 16. Peace and Justice Strong Institutions; 17. Partnerships to Achieve the Goal.[1]

However, the goals lack attention to several significant pieces that move people to the margins. We need to deal with complexity. Within a complex system, we need:

> to map its dynamics; ... locate gateways and leverage points; ...understand how things are changing over time; to create conditions for positive, adaptive latent attractors to emerge and be sustained internally; and to try to undermine negative and destructive attractors.
>
> (Burns and Worsley 2015, p. 37)

FBDOs should work with the poorest to understand the overlapping aspects of marginalization. Rather than say, "aim for universal education" (a laudable goal), I'd say "aim to work with the poorest communities to achieve their goals". The important aspects here will be to listen to marginalized people describe: (1) How they are marginalized and (2) How they can emerge from the margins.

We will begin with two of the aspects that are reflected in the sustainable development goals. Then we move on to several missing pieces.

Environmental poverty

Some people lack a safe and sustainable environment. Living on a garbage dump is one extreme example. Others include pollution, lack of access to clean water, lack of land to grow food. A local community may suffer environmental poverty or some in the community may have access to safer environmental spaces than others. While we begin with the local community determining its needs, the context may quickly grow larger: regional, national, and international. For example, a community may lack water because the water source is controlled by a multinational corporation or a neighboring community upstream may be polluting the water. While many people argue for environmental sustainability, they sometimes fail to consider the structures that prevent sustainability. The SDGs contain several environment-related goals. Each goal is laudable but the local communities must decide for themselves what the priorities are, and have the power to enact them. Further, environmental poverty exists alongside several other pieces, for example, racism.

Indigenous theologies can be helpful here. "A theology that addresses humanity alone and leaves the rest of the cosmos unaddressed is an incomplete theology" (Longchar 2012, p. 96). Environmental marginalization is particularly important to indigenous groups and encompasses the importance of the connection to land.

Since protection of the land is protection of life itself, indigenous people preserve the land and its resources ... *b) By Affirmation of Interconnection ... c) By the Practice of Simple Living ... d) By the Practice of Responsible Ownership.*

(Longchar 2012, pp. 31–32)

First, peoples live in concert with their environment, aiming for balance, and can provide a model for interaction with our environment.

Sexism

The SDGs include references to gender but each target only pertains to women. Females do suffer particular harms. Sexual violence is one example, among many others. Reproductive issues also disproportionately affect females, particularly females in poor countries. Haynes notes that about half a million women die each year from complications of bearing children, 99% in the global South. Another 300,000 are injured or sick from bearing children. Further each year, almost 800,000 babies either are born dead or die within their first week. Haynes cites a World Bank study in Bangladesh, which found that children under ten were far more likely to die if the mother dies (Haynes 2007, pp. 156–7). When a woman dies, her family, particularly her children, are also more likely to remain poor.

However, sexism is broader than attention to women. Much development literature and theology both tend to conflate gender and sex. It is true that women face marginalization and exclusion on the basis of sex and gender; however, sex and gender need to be addressed separately.

The common distinction of gender and sex as masculine/feminine and male/female also excludes many people who are on the spectrum rather than at one end. Intersex people combine aspects of male and female. Transgender people identify more with another sex than their own biological sex. Others identify as neither male nor female. People are discriminated against because of their biological sex or chosen gender attributes. In some societies, people can both be revered and disparaged for their status.

It is not enough to include women, but fail to address masculinities, or ignore people outside the binaries. Often, we ask how we can include women rather than focusing on the deeper issue of the masculinity that can harm and marginalize men, women, and transpeople. Many men do not want to adhere to their traditional gender characteristics of hiding emotions, always being strong, competitive, and so forth. Less masculine men have been degraded. While one can state that with patriarchy women have been treated as less than men, attention to gender and sex is not this simple. Women experience this "less than" in different ways, depending on race, class, religion, and so forth. Men and transpeople can also experience this "less than."

Further, Western concepts of sex and gender may not fit other cultures. Heike Walz, German feminist theologian, uses the phrase "Western gender imperialism."

> In Argentina ... "the" gender perspective had to be implemented as obligatory in every project ... but white people in Europe and Argentina rarely ask the indigenous people about *their* notion of gender, sexuality, womanhood and manhood, the body, and so on.
>
> (Walz 2013, p. 90)

Different cultures have understood sex and gender in a variety of ways. FBDOs need to be attentive to local understandings.

Heterosexism

Sexuality is a missing component from much developmental practice and from the SDGs. However, exclusion due to differing sexualities is common, even in Western societies. There are a variety of sexualities, including but not limited to homosexuality, bisexuality, and heterosexuality. We see this marginalization most clearly in the initial response to HIV/AIDS, and the assumption of its association with non-heterosexual people, but people with differing sexualities face discrimination and marginalization in many countries simply for having a non-normative sexual preference.

In one study, when assessing the stigma associated with HIV/AIDS with sex workers, the development practitioners realized the stigma was about the identities of the workers themselves. By having conversations with the workers, the practitioners found: "one overriding identity... was seen as unacceptable to society to the exclusion of everything else – that of being a sex worker or homosexual or transgendered (Praxis 2013c)" (Narayan, Bharadwaj, and Chandrasekharan 2015, p. 143). In many cases, though one may live at the intersection of privilege and oppression, just one aspect can move a person to the margins. Further, varying aspects of oppression can push one further toward the margins.

Ableism

The SDGs also do not attend to disability directly, another form of marginalization, though they do attend to health. People with disabilities are often marginalized and the poorest are often already living with a disability or at risk of disability. "Hunger ..., lack of clean drinking water, poor sanitary conditions and the toll that hard physical labour takes on the body all predispose the poor to certain categories of illness" (Kabeer 2010, p. 69). The poor are more at risk of acquiring a disability than the rich.

Further ability and disability are relative terms, depending on the environment. "An impaired body is only disabled if the environment is disabling. For example, with contact lenses a person with poor eyesight is not visually disabled" (Walby 2009, p. 270). Our communities determine whether we can be included with a few modifications. We need to modify our spaces to be inclusive. The WHO states 15% of the population has a disability. Rather than create spaces that assume humans are perfect, we can create spaces that enable humans to be fully human.

As Alghaib argues, people with disabilities were either seen as ill and needing a cure or as objects of charity (Alghaib 2015, p. 116). The WHO has begun to shift this limited understanding stating that "'people with disabilities should not be considered 'objects' to be managed, but 'subjects' deserving of equal respect and enjoyment of human rights' (WHO, 2011, p. 34)" (Alghaib 2015, p. 117). Regardless of ability, each human being is fully human. Humans should be equally valued as human.

Ageism (youth and elderly)

While the SDGs address education, they do not focus on the marginalization of youth or elderly people. Often, we exclude the young and the old and focus on people of working age: 18–65. However, this exclusion is inappropriate. In many African countries, for example, more than half the population is under 18. One third of the world's population is under 18. Hence, both education and including youth in all planning and practice are important. In other countries, the elderly have become a larger part of the population at the same time as social safety nets are contracting. We need to ensure that people of all ages can live a full life.

First, we need to include children in development practices. Vicky Johnson, U.K. anthropologist, explains how to engage children in decision-making processes (2015). Children's own experiences can be valued. We need to realize that children are far more capable than we assume. Children should be included in conversations about development, at local, regional, national, and international levels. "Participation transforms the power relations between children and adults, challenges authoritarian structures and affirms children's capacity to influence families, communities and institutions" (157). Joachim Theis, German theologian and child advocate, and Clare O'Kane, U.K. child rights consultant, offer detailed suggestions of how to include children at all levels of planning and practice (Theis and O'Kane 2005).

Second, we need to ensure appropriate education. To survive, each individual needs to gain the skills to do so. It is also important to learn about one's place in the wider world. What education should include cannot be determined in advance by people "bringing development." Instead, locals should be involved in planning and assessing the curriculum. Globally, education should be responsive to local needs. An education that helps us to become fully human needs to begin from the local situation and build out.

In many cases, the education system needs change. Nyambura Njoroge, Kenyan theologian, argues in the Kenyan context that:

> It is vital that we critically evaluate the ... quality of what is taught at every level to ensure that our youth are equipped with the right tools, ones that are self-empowering and self-affirming. ... We must engage in rigorous decolonizing and depatriarchalizing processes.
>
> (Njoroge 2009, p. 178)

A Kenyan education system if set up by outsiders will be inappropriate for Kenyans. Further, attention needs to be paid to the most marginalized.

Freire focused throughout his life on articulating an empowering education. He suggests shifting priorities so that schools "respect the "ways of being" of students," continuously train teachers in praxis, and become creative places. These priorities also entail changing assessment and pedagogy to be student centered (Raynolds, Jr. 1993, p. 11). Ensuring that education focuses on students' experiences and connecting these experiences to the wider world is key to both youth and adult education.

Briefly, I want to attend to the marginalization of the elderly population. Since we tend to focus on people's contribution to economies, we then tend to ignore populations over the age of 65, when we assume people retire. First, this excludes older people with low incomes. Second, many people lucky enough to find work are unable to retire, needing to provide for themselves and their families. Third, human beings contribute to society by being human. People of all ages need to be included in communities.

Racism and ethnocentrism

Racism and ethnocentrism are not included in the SDGs. Surprisingly, the issue of race is often hidden in development practice[2], although there is often a racial difference between Northern practitioners and those in the global South experiencing development.[3] Indigenous groups are marginalized in almost every country, including the USA. This lack of attention to race surprises me given our global history of slavery and colonialism.

> While the centuries-long transatlantic slave trade laid the foundation for the current state of poverty in Africa and the African diaspora, the colonial legacy and the postcolonial condition of African people provide the background for the endemic nature of this crisis in Africa today.
>
> (Olupona 2009, p. xii)

The same is true in the Asian and Western Hemisphere context. Racism continues today; its long history extends its tentacles. "Most of the countries whose people have to swallow the humiliating pill of structural adjustment programmes (SAP) imposed by moneylenders, are also predominantly black. The moneylenders are predominantly white" (Oduyoye 2004, p. 51). The powerful tend to be light-skinned. The powerless tend to have darker skin colors.

Considering race must also include addressing our constructions of whiteness and the spectrum of racism. As Emilie Townes, U.S. womanist theologian, argues, when we see race solely as a bias against "darker-skinned peoples," this narrowing:

> invites folks of European descent and others to ignore the social construction of whiteness.
>
> allows darker-skinned racial ethnic groups to ignore their internal color caste system.
>
> often opens the door for weird bifurcations of class, race, gender, age and so on.
>
> (Townes 2013, p. 71)

Racism is experienced differently in differing contexts. In Haiti, a person of African heritage with wealth can be white. In Brazil, there is a racial hierarchy from lighter to darker skin with racism across the spectrum. Close attention to local variations of racism as well as the racial disparity between those "developing" and those "developed" is critical.

Nationalism and xenophobia

Nationalism and xenophobia are missing from the SDGs. Nationalism is seeing other nations as inferior. Xenophobia is a fear/hatred of "foreigners." Both nationalism and xenophobia are common. Elina Hankela, Finnish liberation theologian, argues that in South Africa, "African migrants are commonly perceived as sources of crime, prostitution and disease, and blamed for taking jobs and 'our' women, while their countries of origin have been portrayed as inferior to South Africa" (Hankela 2014, p. 73). In part, racism plays a role. The same connection of race and xenophobia occurs in other contexts. For example, "The Moroccan media was perpetuating fear with articles like Maroc Hebdo's 'Le péril noir' ('The black peril,' Najib 2012), linking the existence of irregular sub-Saharan Africans to acts such as murders, robberies and serious crimes" (Ngo 2018, p. 16). Further examples can be found in the USA and across Europe.

This concept is particularly important to address given the huge numbers of migrants due to economic, environmental, religious, and political disasters and conflicts. Migration has more than doubled in the past 30 years. People can even be "stateless." While two-thirds of migrants move to richer countries, one-third move between poorer countries, and poor countries receive many refugees (Ramachandra 2008, pp. 112–3). I began the book with an example from Jordan, a country which has repeatedly hosted hundreds of thousands of refugees.

We tend to distinguish between "refugees" and "economic migrants" as if these categories are clear. Commonly, we perceive refugees to be people who have to leave their country, fearing for their lives. Economic migrants, in contrast, we assume "choose" to leave in order to find work elsewhere. However, the terminology is determined by governments. Not all countries recognize "refugees" from all areas. "Instead of being regarded as innocent until proven guilty, they are assumed to be fraudulent" (Ramachandra 2008, p. 158). Many people end up in another country without formal documentation. Sometimes this is due to different perceptions of "choice" between the migrant and the government.

Borders are geographical and political. Borders define the limits of a state or region, including some and excluding others. This situation is even more complex in this century. Vinoth Ramachandra, Sri Lankan theologian, explains that migrants arriving in France and Spain seeking refuge are moved to North African countries and placed in "transit centers" while their cases are considered (Ramachandra 2008, p. 158). They are neither in France/Spain nor in the African country.

Further, borders are dangerous places for many migrants. "A migrant told me that no one passes through this border [between Morocco and Algeria] without being violated in some way – financially, physically, sexually or all three" (Ngo 2018, p. 17). Lacking formal documentation pushes people to the margins where they suffer further abuse.

Political poverty

Political poverty is the inability to participate in local, regional, national, and international government systems effectively. Migrants often experience political poverty but so too do people born, raised, and living in one country. Two SDGs partially address the political realm (16 and 17).

A government's role should be to protect and sustain its members; however, many issues expand beyond a government's borders. The political realm also encompasses far more than formal governments. It includes any grouping that governs human relationships. To live in community, we have to find ways to survive and thrive together.

In terms of who should participate, Nancy Fraser, U.S. political theorist, argues for the *"all-subjected principle"* (Fraser 2008, p. 65). If one has to obey rules, s/he should be involved in setting the rules. Further, s/he needs to be equally involved to others. Fraser calls this concept *"parity of participation"* (Fraser 2008, p. 60). Each person should be able to equally participate. Democracy can be more than just access to a free market.[4] Civil society, including FBDOs, has become another space for political action, which is positive; however, it can leave economies and governments untouched.

There should be heterogeneity in the political solutions proposed, acknowledging the diversity within and between communities. For example, "The decision-making process relies, in many non-European cultures, on consensus. It involves a show, careful attempt to safeguard the collective harmony" (Verhelst 1990, p. 40). Democracy can be seen as the dominance of the majority. There are other systems and modifications that can ensure all voices are included in decision-making.

Local, regional, and national governments should

> act for the good of all their citizens: what *Philippines* refers to as "the new patriotic politics, i.e. politics based on a political programme for the good of the majority" (p. 22). … Safety nets should be provided for the most vulnerable.
>
> (Taylor 2003, p. 44)

The political realm can help to address the conflicts that result from race, sex, sexuality, religion, and so forth.

Politics must also move beyond the nation-state, as noted in the xenophobia section. Many people born in one country live in another and many people living in a particular country have ties elsewhere. "The nation-state … has become instead the place of the ethni" (Inda and Rosaldo 2008, pp. 23–4). A nation-state needs to actively include all people within its borders. As diverse communities living together in one state: "The different communities enjoy varying degrees of autonomy, and are encouraged by the state (not least through the media and educational institutions) to enter into a continuous public conversation with one another on issues of mutual concern" (Ramachandra 2008, p. 153). Rather than emphasizing a form of nationalism where all are similar, the state can be a space where diverse groups learn to live together.

Religious discrimination

Religion is not addressed in the SDGs. One aspect within FBDOs is the prioritization of people within one's own religious tradition. Another is the aim to convert to one's own tradition. Both are wrongheaded and barriers to development. Mercy Oduyoye, Ghanian Methodist theologian, calls this "religious chauvinism" (2004, p. 51). We need to engage with the diversity of beliefs and traditions. "Sharing spirituality across religious boundaries will make us neighbours who honour each other's specificities while at the same time seeking mutual caring and sharing and learning together" (Oduyoye 2004, p. 51). Some religions end their respect for diversity at the door of their own religion.

In an earlier chapter, I noted the "good news" within the Christian tradition as being able to work toward the new heaven and new earth and living abundant life. Some strands of the Christian tradition narrowed this to a solely spiritual salvation. Instead, the Christian mission is to walk with the poor and struggle with them. We need to remember that the good news is the ability to live life abundantly.

"Justice demands that we respect the other (individual and collective) in his/her/their 'otherness' – which also includes religion" (Wilfred 2007, p. 146). In living life abundantly, we need to respect and engage with the variety of ways to know God.

Dialoguing between and within religious traditions is crucial to living life abundantly. Wendy Tyndale, U.K. development scholar and practitioner, provides one example from Guatemala where Presbyterian, Catholic, and Mayan women "cultivate flowers and strawberries for the local market. ...They all believe that the Creator – whether experienced as the Trinitarian God or as the Heart of the Earth and the Heart of the Sky – has given them life" (Tyndale 2005, p. 159). There are several varieties of indigenous theologies around the globe that could be particularly helpful here: for example, in Chiapas (Marcos 2013), in India (Longchar 2012, 2013), and in North America (Tinker 2013). The key is to be attentive to local theologies and religious expressions. It is crucial to understand them to sensitively approach local issues, particularly where there are already conflicts between local religious groups.

Examples of working with the most marginalized at their intersections

Partners in Health (PIH), while not an FBDO, follows notions of empowerment, justice, and working with the most marginalized. "For Partners in Health, the goal in places like Haiti is to accompany people in their suffering and to seek the changes necessary to remedy their suffering and the entire situation that has given rise to their suffering" (Gutiérrez 2013, pp. 30–1). PIH's aim is first to reduce premature deaths. One example of success in this area, according to Farmer is Rwanda, where AIDS, malaria, and TB are causing far fewer deaths. The national health system has worked to include all Rwandans, particularly the rural poor. Further, it is improving its capability to provide services to all. Farmer notes that Rwanda is focused on training health care professionals, in particular. "Long experience suggests that strong public institutions are a prerequisite of meaningful gains in these areas. Those who work with nongovernmental organizations and universities and churches are by definition outside of these public institutions" (Farmer 2013a, p. 124). One of the keys for PIH is to work with the national health care system to improve it, rather than outside, so that change is sustainable beyond NGO involvement.

For PIH, the poorest deserve the best health care. Rather than using the leftovers for the poor, "Whenever medicine seeks to reserve its finest services for the destitute sick, you can be sure that it is option-for-the-poor medicine" (Farmer 2013b, p. 56). Health care should not be a commodity, bought and sold. The poorest rely most on their health and deserve the best care.

While this particular organization is not faith-based, PIH is a critical example for FBDOs. Faith-based health care accounts "for perhaps 30–50 percent of the global health care landscape" (Marshall and Van Saanen 2007, p. 33). Hence, the perspective on who should get what care is critical. Another issue within faith-based health care is whether the organizations train local people to be health care professionals and work to improve the national health care system. Delivering health care without enabling local and national communities to sustainably develop their own health care system perpetuates injustice.

When development organizations listen to the marginalized and work at the intersections, they enact justice. Another example is the work of Bishop Verryn and the Anglican Church in Johannesburg with migrants to South Africa, excellently analyzed in Hankela's book *Ubuntu, Migration and Ministry*. Similarly, the work of a Protestant Church in Morocco with sub-Saharan African migrants is detailed by Ngo in her book *Between Humanitarianism and Evangelism in Faith-based Organizations*. Both organizations deal with the intersections of economic and political poverty, xenophobia, and other areas. These local efforts should be accompanied by national and international advocacy.

One further example emerged in Brazil in the 1980s, the Movimento dos Trabalhadores Rurais sem Terra (MST – Brazilian Landless Movement).[5] With the support of some of the Catholic Church, landless peasants moved alongside[6] and farmed unused but arable land and attempted to gain cooperative ownership. The MST argues that the landless need access to land to sustain life. Brazil's land distribution is extremely unequal, stemming all the way back to the original splitting of the land into 15 parcels to encourage a dozen Portuguese men to invest in Brazil. There are still farms the size of small U.S. states in Brazil today and much of the land lies unused.

The MST gathers cooperatives of people (on average 250 families) who farm a piece of land for at least one year and then apply for ownership through the courts.[7] The Brazilian Constitution, written after the return to democratic rule in 1984, stated that to keep ownership of arable land, the land must be used productively. This shift opened up the possibility for unused land to be taken over by the landless. The MST has helped more than 370,000 families resettle into cooperatives on unused land. There are hundreds of thousands of people trying to gain access to land and about 150,000 families currently farming land they hope to own one day. The successful settlements focus on sustainability, include community health care, education, and so forth, and include children in decision-making processes.

Finally, there are several examples of Praxis Institute for Participatory Practices[8] working at the margins effectively. One success was rebuilding on Vilufushi in the Maldives after the 2004 tsunami, when the geographical size of this island tripled.

> "It was decided to empower the island community itself to collectively take the decision [to identify beneficiaries]. ... The results challenge the conventional narrative that in a post-disaster context, affected communities are too fragmented by conflicts of interest to be able to find common agreement."
>
> (Joseph, Kisana, and George 2015, p. 25)

Even in emergencies, people can and want to participate in their own recovery.

Conclusion

We can see from these examples that it is possible to work with the poorest communities at the intersections of their marginalization. Rather than choose "a project," FBDOs need to commit to walking alongside the poorest communities as they work for change. FBDOs also need to amplify the voices of the poorest at national and international levels. Finally, FBDOs can also examine the intersections within their own organizations to include those who may be excluded.

FBDO Action:

1 Be with the poorest. Walk with the poorest.
2 Amplify the intersections the marginalized articulate to the powerful.
3 Articulate how to change structures to remove these barriers.

Notes

1 See www.un.org/sustainabledevelopment/sustainable-development-goals/ for details of each.
2 One exception to this rule was in the South African context, where Muslims also contended with race.

> South African Muslims engaged in the anti-apartheid struggle routinely came together in religious circles (*halaqat*) to reflect collectively on a translation of the Qur'an, asking on another what they felt the various verses meant and how these verses spoke to their experiences.
>
> (Rahemtulla 2017, p. 18)

3 I have found only two major exceptions to this silence: Kalpana Wilson (2012) and Uma Kothari (2006a, 2006b).

4 Walby notes that few countries, if any, are fully democratic.

> Ten indicators of the depth of democracy in a country are: 1. No hereditary or unelected positions…; 2. No colonies …; 3. No powers of governance held by an additional non-democratic polity …; 4. Universal suffrage…; 5. Elections, especially those that are free, fair, and competitive…; 6. A low cost for electioneering…; 7. An electoral system with proportional representation; 8. An electoral system with quotas for under-represented groups such as women; 9. A proportionate presence in parliament of women and minorities; 10. A range of institutions (e.g. welfare services) that are governed by the democratic polity.
>
> (Walby 2009, 180)

5 See www.mst.org for further details. Or for an English website: www.mst-brazil.org.

6 Groups cannot actually live on the land until they gain ownership of it; they can, however, farm it. Hence, you see temporary settlements of tents and tarps along roadsides and in ditches.

7 Although one can apply for the title after one year, on average a case takes several years to get through the court system.

8 See www.praxisindia.org.

References

Alghaib, Ola Abu. 2015. Aid Bureaucracy and Support for Disabled People's Organizations: A Fairy Tale of Self-Determination and Self-Advocacy. In *The Politics of Evidence and Results in International Development: Playing the Game to Change the Rules?*, eds. Rosalind Eyben, Irene Guijt, Chris Roche, and Cathy Shutt, 115–33. Bourton on Dunsmore: Practical Action. doi:10.3362/9781780448855.007

Bailey, Barbara. 2009. Feminization of Poverty across Pan-African Societies: The Church's Response-Alleviative or Emancipatory? In *Religion and Poverty: Pan-African Perspectives*, ed. Peter Paris. Durham: Duke University Press, 39–65. doi:10.1215/9780822392309-003

Burns, Danny and Stuart Worsley. 2015. *Navigating Complexity in International Development: Facilitating Sustainable Change at Scale*. Bourton on Dusmore: Practical Action. doi:10.3362/9781780448510

Chambers, Robert. 2012. *Provocations for Development*. Bourton on Dunsmore: Practical Action. doi:10.3362/9781780447247

Farmer, Paul. 2013a. Conversion in the Time of Cholera: A Reflection on Structural Violence and Social Change. In *In the Company of the*

Poor: Conversations with Dr. Paul Farmer and Fr. Gustavo Gutiérrez, eds. Michael Griffin and Jennie Weiss Block, 95–145. Maryknoll: Orbis Books.

———. 2013b. Health, Healing, and Social Justice: Insights from Liberation Theology. In *In the Company of the Poor: Conversations with Dr. Paul Farmer and Fr. Gustavo Gutiérrez*, eds. Michael Griffin and Jennie Weiss Block, 35–70. Maryknoll: Orbis Books.

Fraser, Nancy. 2008. *Scales of Justice: Reimagining Political Space in a Globalizing World*. New York: Columbia University Press. doi:10.1111/j.1467-8675.2012.00674.x

Groody, Daniel, interviewer. 2013. Reimagining Accompaniment: An Interview with Paul Farmer and Gustavo Gutiérrez. In *In the Company of the Poor: Conversations with Dr. Paul Farmer and Fr. Gustavo Gutiérrez*, eds. Michael Griffin and Jennie Weiss Block, 161–88. Maryknoll: Orbis Books.

Gutiérrez, Gustavo. 2013. Saying and Showing to the Poor: "God Loves You". In *In the Company of the Poor: Conversations with Dr. Paul Farmer and Fr. Gustavo Gutiérrez*, eds. Michael Griffin and Jennie Weiss Block, 27–34. Maryknoll: Orbis Books.

Haynes, Jeffrey. 2007. *Religion and Development: Conflict or Cooperation?* New York: Palgrave. doi:10.1057/9780230589568

Hankela, Elina. 2014. *Ubuntu, Migration and Ministry: Being Human in a Johannesburg Church*. Boston: Brill. doi:10.1163/18748945-02901019

Inda, Jonathan Xavier and Renato Rosaldo. 2008. Tracking Global Flows. In *The Anthropology of Globalization: A Reader*, eds. Jonathan Xavier Inda and Renato Rosaldo, 2nd ed., 3–46. Malden: Blackwell.

Johnson, Vicky. 2015. Valuing Children's Knowledge: The Politics of Listening. In *The Politics of Evidence and Results in International Development: Playing the Game to Change the Rules?*, eds. Rosalind Eyben, Irene Guijt, Chris Roche, and Cathy Shutt, 155–71. Bourton on Dunsmore: Practical Action. doi:10.3362/9781780448855.009

Joseph, M.J., Ravikant Kisana, and Mary George. 2015. Building Consensus Methodically: Community Rebuilding in the Maldives. In *Participation Pays: Pathways for post-2015*, eds. Tom Thomas and Pradeep Narayanan, 25–39. Bourton on Dunsmore: Practical Action. doi:10.3362/9781780448695.003

Kabeer, Naila. 2010. Alternative Accounts of Chronic Disadvantage: Income Deficits Versus Food Security. In *What Works for the Poorest? Poverty Reduction Programmes for the World's Extreme Poor*, eds. David Lawson, David Hulme, Imran Matin, and Karen Moore. Bourton on Dunsmore: Practical Action.

Kothari, Uma. 2006a. An Agenda for Thinking about 'Race' in Development, *Progress in Development Studies*, vol. 6:1, 9–23. doi:10.1191/1464993406ps124oa

———. 2006b. Critiquing 'Race' and Racism in Development Discourse and Practice. *Progress in Development Studies*, vol. 6:1, 1–7. doi: 10.1191/1464993406ps123ed

Longchar, Wati. 2012. *Returning to Mother Earth: Theology, Christian Witness, and Theological Education: An Indigenous Perspective*. Kway Jen: Programme for Theology and Cultures in Asia.

_____. 2013. Liberation Theology and Indigenous People. In *The Reemergence of Liberation Theologies: Models for the Twenty-first Century*, ed. Thia Cooper, pp. 111–21. New York: Palgrave.

Mander, Harsh. 2015. *Looking Away: Inequality, Prejudice and Indifference in New India*. New Delhi: Speaking Tiger.

Marcos, Sylvia. 2013. Embodied Theology: Indigenous Wisdom as Liberation. In *The Reemergence of Liberation Theologies: Models for the Twenty-First Century*, ed. Thia Cooper, pp. 123–33. New York: Palgrave.

Marshall, Katherine and Marisa Van Saanen. 2007. *Development and Faith: Where Mind, Heart, and Soul Work Together.* Washington, DC: World Bank. doi:10.1596/978-0-8213-7173-2

Narayanan, Pradeep, Sowmyaa Bharadwaj, and Anusha Chandrasekharan. 2015. Re-Imagining Development: Marginalized People and the Post-2015 Agenda. In *Participation Pays: Pathways for Post-2015*, eds. Tom Thomas and Pradeep Narayanan, 137–55. Bourton on Dunsmore: Practical Action. doi:10.3362/9781780448695.009

Ngo, May. 2018. *Between Humanitarianism and Evangelism in Faith-based Organisations: A Case from the African Migration Route*. London: Routledge. doi:10.4324/9781315561479

Njoroge, Nyambura. 2009. The Struggle for Full Humanity in Poverty-Stricken Kenya. In *Religion and Poverty: Pan-African Perspectives*, ed. Peter Paris. Durham: Duke University Press, 166–90. doi:10.1215/9780822392309-009

Oduyoye, Mercy. 2004. *Beads and Strands: Reflections of an African Woman on Christianity in Africa*. Maryknoll: Orbis.

Olupona, Jacob. 2009. Foreword: Understanding Poverty and Its Alleviation in Africa and the African Diaspora. In *Religion and Poverty: Pan-African Perspectives*, ed. Peter Paris, ix–xx. Durham: Duke University Press.

Rahemtulla, Shadaab. 2017. *Qur'an of the Oppressed: Liberation Theology and Gender Justice in Islam*. doi:10.1093/acprof:oso/9780198796480.001.000

Ramachandra, Vinoth. 2008. *Subverting Global Myths: Theology and the Public Issues Shaping Our World*. Downers Grove: IVP Academic. doi:10.1558/poth.v13i4.520

Raynolds, Jr., Harold. 1993. Preface. In *Pedagogy of the City*, Paulo Freire. New York: Continuum.

Taylor, Michael. 2003. *Christianity, Poverty and Wealth: The Findings of 'Project 21'*. London: SPCK.

Theis, Joachim and Claire O'Kane. 2005. Children's Participation, Civil Rights and Power. In *Reinventing Development? Translating Rights-Based Approaches from Theory into Practice*, eds. Paul Gready and Jonathan Ensor, 156–70. London: Zed Books.

Tinker, Tink. 2013. American Indian Liberation: Paddling a Canoe Upstream. In *The Reemergence of Liberation Theologies: Models for the Twenty-first Century*, ed. Thia Cooper, 57–67. New York: Palgrave.

Townes, Emilie. 2013. Uninterrogated Coloredness and Its Kin. In *The Reemergence of Liberation Theologies: Models for the Twenty-First Century*, ed. Thia Cooper, 69–74. New York: Palgrave. doi:10.1057/9781137311825_9

Verhelst, Thierry. 1990. *No Life without Roots: Culture and Development*. Translated by Bob Cumming. London: Zed Books.

Walby, Sylvia. 2009. *Globalization & Inequalities: Complexity and Contested Modernities*. London: Sage. doi:10.4135/9781446269145

Walz, Heike. 2013. Key Issues for Liberation Theology Today: Intercultural Gender Theology, Controversial Dialogues on Gender and Theology between Women and Men, and Human Rights. In *The Reemergence of Liberation Theologies: Models for the Twenty-First Century*, ed. Thia Cooper, 89–99. New York: Palgrave. doi:10.1057/9781137311825_11

Wilfred, Felix. 2007. Indian Theologies: Retrospect and Prospects, a Sociopolitical Perspective. In *Another Possible World*, eds. Marcella Althaus-Reid, Ivan Petrella and Luiz Carlos Susin, 131–61. London: SCM Press.

Wilson, Kalpana. 2012. *Race, Racism and Development: Interrogating History, Discourse and Practice*. London: Zed Books.

Conclusion

Christian FBDOs need to attend to and work at the intersections of marginalization: environment, sex, gender, sexuality, ability, age, race, nationality, politics, religion, and so forth. In terms of economics, we need to address the assumptions of capitalism that lead to injustice, end the focus on economic growth, and advocate for alternatives that work toward justice. Most importantly, development institutions and practitioners should shift from a focus on development projects to walking alongside the poorest and amplifying the voices of the poorest to the powerful. Practitioners need to become facilitators and mediators.

These shifts come from Christian understandings of empowerment, justice, and being on the side of the marginalized. Jesus walked with the marginalized and brought the good news that humans could live life abundantly. While the church, post-Constantine, changed the understanding of the good news, we can remember the original mission and accompany the marginalized. Similarly, the biblical texts urged us to work toward justice, giving away all but what was needed, and the early church followed this urging, understanding one could only enact charity if a just situation already existed. Today, we have shifted the notion of charity to giving to the poor as optional and narrowed the notion of justice to commutative. Instead, we can return to seeing ourselves in communion with others, free to work toward the common good, free to desire abundant life with God and others, understanding abundance and the concept of "enough," and ensuring we treat each human being as fully human. While we participate in structures that exercise power over, we can see in the life of Jesus and early Christianity that God showed a different form of power: empowerment. For each person to live abundant life, s/he needs to be empowered: having power to, power with, and power within.

Christian FBDOs should work toward empowerment of the marginalized and accompany the marginalized. In the North, FBDOs should advocate for changes to structures of domination, articulating changes toward empowerment, and a holistic understanding of justice. Returning to biblical and early church notions of empowerment and justice can help us to shift our development practices. This shift is critical as the majority of Christians are people who experience development.

Let us return to the questions that first emerged as Christians addressed development: (1) Is development salvation? (2) What should the North do with its riches? (3) What do the new heaven and new earth mean for structures? And (4) Should humanity be at the center of development?

First, development can be part of the work of salvation if we convert to the poor, walk alongside the poorest and most marginalized, enable each human being to be empowered, and work toward justice in all realms. Here, sin is understood as injustice, that which separates humans from each other and God. Salvation is understood as conversion to the side of the poor and working toward the new heaven and new earth where justice will rule.

Second, the wealthy (including structures), wherever they reside in the globe, need to redistribute wealth to the poorest. We need to advocate for an end to unlimited accumulation. We can begin by ending loans to the global South, as the money already belongs to the poorest and should not further enrich the global North. Our understanding of poverty also needs to expand to include all the ways human beings are marginalized. Second, our understanding of wealth needs to change from unlimited accumulation to living life abundant. Development (evolving and changing) need not equal growth (getting bigger).

Third, working toward the new heaven and new earth means resisting structures of domination and advocating for structures that empower the poorest and most marginalized, and work toward a holistic notion of justice in all realms. Too often we accept structures as a given; instead we can propose and work toward alternatives.

Fourth, yes, each human being deserves to live life abundant. Any development practice that excludes or marginalizes a human being fails to be a Christian development practice, whether it be economic, political, religious, racial, and so forth.

In sum, FBDOs need to address the intersections of marginalization, stop the focus on economic growth, and let the poorest decide how to improve their own lives. Because Jesus resides at the margins, we should be there too working to do justice, focused on empowerment rather than domination.

While this book articulates a theology of international development, I want to make clear that marginalization and injustice occur within every country in the world. During the research for this book, I happened to read at the same time two books that strikingly brought this reality to my attention. Mark Schuller, U.S. anthropologist, and Pablo Morales, journalist, edited a book, which articulates the huge inequalities in recovery after the earthquake in Haiti (January 2010) while Vincanne Adams, U.S. anthropologist, articulates the huge inequalities in recovery after the Hurricane Katrina in the USA (August 2005). (Schuller and Morales 2012; Adams 2013) I suggest reading these two books side by side and then decide whether the situation of the poorest is somehow better with a U.S. response than a Haitian response. The combination of a focus on the business and NGO realms along with the relative sidelining of direct government help leads to stunningly inappropriate treatment of the poorest in both situations.

This book offers starting points for theology and practices. I hope FBDOs will take these concepts forward. I also hope other scholars will take these themes forward; in particular, I look forward to more theology from the global South, advocating for and challenging the interpretations of the good news.

References

Adams, Vincanne. 2013. *Markets of Sorrow, Labors of Faith: New Orleans in the Wake of Katrina*. Durham: Duke University Press.

Schuller, Mark and Pablo Morales, eds. 2012. *Tectonic Shifts: Haiti Since the Earthquake*. Sterling: Kumarian Press.

Index

Note: Page numbers followed by "*n*" refer to notes.

Printed in the United States
by Baker & Taylor Publisher Services

Printed in the United States
by Baker & Taylor Publisher Services